Artificial Intelligence for Climate Hazards

Adam Jabłoński
Marek Jabłoński

CRC Press
Taylor & Francis Group
Boca Raton London New York

CRC Press is an imprint of the
Taylor & Francis Group, an **informa** business
A SCIENCE PUBLISHERS BOOK

First edition published 2026
by CRC Press
2385 NW Executive Center Drive, Suite 320, Boca Raton FL 33431

and by CRC Press
4 Park Square, Milton Park, Abingdon, Oxon, OX14 4RN

CRC Press is an imprint of Taylor & Francis Group, LLC

Library of Congress Cataloging-in-Publication Data (applied for)

ISBN: 978-1-032-56129-5 (hbk)
ISBN: 978-1-032-56130-1 (pbk)
ISBN: 978-1-003-43401-6 (ebk)

DOI: 10.1201/9781003434016

Typeset in Palatino Linotype
by Prime Publishing Services

This book is dedicated to Professor Henryk Arctowski and Professor Antoni Bolesław Dobrowolski, Polish polar explorers and members of the famous Belgica scientific and research expedition, considered the world's first climate scientists

Taylor & Francis
Taylor & Francis Group
http://taylorandfrancis.com

Preface

The aim of the monograph is to present a modern and original approach to the design and implementation of effective and efficient artificial intelligence solutions that ensure the induction of positive climate change. The monograph presents the application of digital business models and strategies combining solutions for digital and climate transformation. Digital platform solutions based on algorithms changing the functioning of technical organizations will be presented. Applied technological management, supported by elements of automation and robotics, will shape the emergence of new climate services. The mechanisms for implementing cybersecurity and safety principles described herein will serve to reduce climate hazards.

The aim of the book is to identify assumptions for the use of artificial intelligence in monitoring climate change.

The problem thus defined creates opportunities for the creation of theoretical and practical models outlining the mechanisms of using artificial intelligence solutions in climate change monitoring and management.

The book concerns specialist knowledge in the following fields:

- Artificial intelligence as a research object in management sciences and climate sciences and as a practical tool for modern business solutions.
- Systematic literature review – a description of the landscape of AI research.
- Artificial intelligence – key meanings and interpretations in the context of climate change.
- Artificial intelligence and an overview of selected solutions for creating positive climate changes.
- Climate hazard management mechanisms using artificial intelligence solutions.
- Contextualization of the place and role of artificial intelligence in climate hazard management.

- The authors' original concept for the development of the evolutionary cycle of climate business models – from sustainable business models to AI Climate-Changing Business Models.
- The authors' original concept of climate hazard management using artificial intelligence.
- The authors' original concept of developing the mechanisms and solutions of artificial intelligence in the management of climate hazards.
- The authors' original spiral of knowledge of the use of artificial intelligence mechanisms to solve problems in the field of climate hazards and climate change.

In order to illustrate the complexity of the issue presented in the monograph, the components that are chapters within the scope of the authors' original concept of artificial intelligence in the fight against climate hazards have been presented graphically in Figure 1.

The presented model identified a closed cycle of the implementation of AI principles in the fight against climate hazards. This proposed model is described in detail in this scientific monograph. The aim of the monograph is to indicate the theoretical and practical determinants of artificial intelligence in the fight against climate hazards. The scientific discussion in this monograph concerns the solution to the scientific problem concerning the impact of artificial intelligence solutions on climate change monitoring and the search for the impact of business and social solutions on its neutralization.

The book is preceded by this preface. Chapter one describes the main purpose of the scientific discussion.

Chapter two refers to artificial intelligence as a research object in management and climate sciences and as a practical tool for modern business solutions. A systematic literature review – a description of the landscape of AI research – is described in chapter three. Chapter four discusses artificial intelligence – key meanings and interpretations in the context of climate change.

Chapter five is dedicated to the important issue of artificial intelligence and an overview of selected solutions for creating positive changes for the climate. Chapter six deals with climate hazard management mechanisms using artificial intelligence solutions. Chapter seven contextualizes the place and role of artificial intelligence in managing climate hazards. Chapter eight presents the authors' original concept of the development of the evolutionary cycle of climate business models – from sustainable business models to AI Climate-Changing Business Models. The authors' original concept of climate hazard management using artificial intelligence is included in chapter nine. The authors' original concept of developing

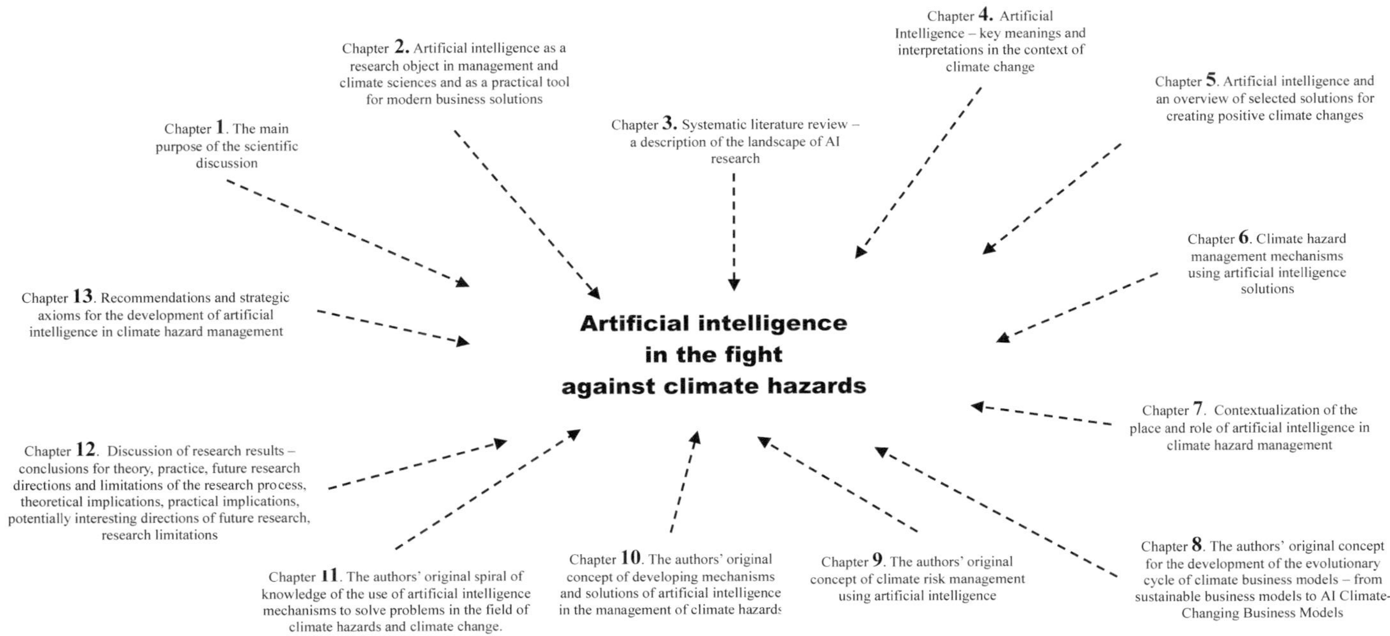

Figure 1. Artificial intelligence in the fight against climate hazards Source: Own study.

mechanisms and solutions of artificial intelligence in the management of climate hazards is described in chapter ten. Chapter eleven contains the authors' original spiral of knowledge of the use of artificial intelligence mechanisms to solve problems in the field of climate hazards and climate change. The discussion of research results – conclusions for theory, practice, future research directions and limitations of the research process, theoretical implications, practical implications, potentially interesting directions of future research, and research limitations – are included in chapter twelve. Chapter thirteen, the last one, presents recommendations and strategic axioms in the field of developing artificial intelligence in the management of climate hazards. The monograph ends with the conclusion, a list of figures and tables.

Adam Jabłoński
Marek Jabłoński

Contents

List of Figures

List of Tables

Table 2. Diagram of the process of the research procedure in empirical sciences and the process of diagnostic procedure in psychology developed by Spendel, Z. (2005), interpreted for the purpose of conducting advanced research in the area of artificial intelligence for climate hazards.

Scientific research – classical	**Scientific study – artificial intelligence for climate hazards**		**Diagnostic procedure – classical**	**Diagnostic procedure – artificial intelligence for climate hazards**
Recognizing and verbalizing the research problem concerning the relationship between selected phenomena, formulating a preliminary hypothesis and (possibly) showing the circumstances of the problem.	The recognition and verbalization of the research problem results from the current needs of the manager and he/she is the generator of the research problem. The research problem is a reflection of a real managerial problem in a business run by a manager. He/she can independently conduct the research process with regard to the research problem posed and/or use the help of a professional researcher. The circumstances of addressing a given research problem are the specific, actual and current context of the organization in a given socio-economic ecosystem in which the company exists.	**Conceptualization**	Preliminary identification and assessment of the problem and justification for addressing it – usually it means finding some irregularities.	A problem is recognized and verbalized as part of a diagnostic procedure only by a manager who needs to solve a given diagnostic problem. The manager, noticing some problems that occur in the organization, looks for effective and efficient ways to solve them.
Positioning the problem and hypothesis in a theoretical context, defining the assumptions and formulating research subhypotheses in terms of relationships between the separated variables.	The manager and/or the researcher who supports him/her quantifies and specifies his/her problem. They are an input to the specification of the adopted research problem, which is in fact a business problem because it directly results from the business needs of a specific company.	**Specification**	Formulation of detailed questions and diagnostic hypotheses on the genesis and pathomechanism of the identified abnormality.	The manager develops a problem that bothers him/her in detail, which he/she must solve. For this purpose, it generates research questions or hypotheses.

Determination of the criteria of empirical applicability and quantitative recognition for observational terms occurring in subhypotheses.	The manager and/or the researcher assisting him/her develops an execution framework in terms of the algorithm of solving a given problem. He/she can use artificial intelligence solutions here.	**Operationalization**	Planning (or selecting an appropriate model) of the diagnostic procedure and the selection of suitable (i.e., relevant to the hypotheses) diagnostic methods.	The manager prepares a list of specific steps along with the proposed diagnostic methods.
Selection of models for collecting and organizing experience compared with the content of empirical implications hypotheses and models of data analysis. Application of selected research and statistical models of selected research and statistical models.	The manager and/or the researcher assisting him/her makes a constructive comparison of the principles of experience management in the organization and the results of evidence research, comparing the results of resource management in a given organization with the research problem to be solved.	**Comparison**	Collection of data by means of selected methods and analysis of results in terms of their compliance with diagnostic hypotheses.	The manager aggregates the obtained results and compares them with a problem in the area of diagnostic procedures, verifying them through previous decision-making processes in the same or a similar area of business.
Formulation of conclusions on the veracity of the tested hypotheses, as well as the indication of the acceptable reference range of the obtained research results to: (1) assumptions and research hypotheses, (2) the possibility of formulating further problems, possibly proposing theoretical and methodological postulates, revising the current paradigm, and (3) taking practical actions on that basis.	The manager and/or researcher assisting him/her will develop a set of goals to achieve research results. It takes into account, in the form of possible micro-corrections, information, data and knowledge from other sources supporting the decision-making processes.	**Application**	Formulation of the diagnosis (in the form of conclusions, opinions, rulings, etc.) and postulates regarding corrective, therapeutic conduct and cooperation with other specialists and institutions.	The manager interprets, diagnoses and synthesizes the results obtained as a result of the procedure, correcting them, if necessary, for other resulting sources that may correct potential errors in the decision-making diagnostic process.

Source: Own study based on Spendel (2005, p. 33).

Cognitive objectives

Cognitive objectives include:

1. Identifying factors responsible for shaping artificial intelligence in the fight against climate hazards.
2. Determining the relationships between AI solutions and climate hazards.
3. Assessing the impact of artificial intelligence in the fight against climate hazards on the effectiveness and efficiency of organizational management.
4. Assessing the relationship between AI solutions and the ability to monitor and influence positive climate change by creating business ecosystems that change the approach to the global economy. This includes the neutralization of large industry into fragmented business models providing the same industrial effect, but without excessive pollution of the natural environment.

Methodological objectives

Methodological objectives include:

1. Indicating the configuration of factors that shape the principles of artificial intelligence development in the fight against climate hazards.
2. Developing strategic recommendations for the development of artificial intelligence in the fight against climate hazards.

Utilitarian objectives

Utilitarian objectives include:

1. Developing the author's original methodology for shaping the principles of artificial intelligence development in the fight against climate hazards.
2. Developing a set of management, technical and operational factors which have the greatest impact on creating the principles of artificial intelligence in the fight against climate hazards.

Key research questions

The following research questions were posed within the framework of the defined research objectives:

1. What methods in the field of management, economic decision analysis systems and improvement of technical efficiency affect the processes of artificial intelligence development in the fight against climate hazards?

2. What concepts developed in the course of scientific research can support the processes of reducing the impact of climate change on people's quality of life and the creation of new business models?

Embedding the topic in scientific theories:

Due to the multidimensionality and holistic nature of the functioning of companies, the topic of the scientific discussion requires one to address issues related to the theory of management and quality science, economics and finance, climate science and technical sciences.

This book is based on the following limitations:

1. So far, there has been no such extensive research into artificial intelligence in the fight against climate hazards and climate change. No significant experience in this regard exists.
2. There is a lack of adopted and recognized AI solutions that would be most effective in eliminating and neutralizing climate hazards.
3. There is an underutilized group of emerging new climate hazards which relevant AI solutions need to be matched to.

Time relevance, the importance of issues and the contribution to science of issues related to the mechanisms of artificial intelligence in the fight against climate hazards.

By writing this monograph, the authors wished to bring added value to the identification, conceptualization and operationalization of the implementation of the principles of artificial intelligence development in the fight against climate hazards. The proposed research issue is generally not presented with such a broad approach in the relevant literature. The value of the creative contribution is related to the presentation of a comprehensive approach to the development of artificial intelligence in the fight against climate hazards. The authors would like to draw attention to the need to develop a new theory of building and implementing the principles of artificial intelligence development in the fight against climate hazards and climate change and to indicate utilitarian solutions that provide guidance for managers of various types of companies interested in furthering the principles of artificial intelligence development in the fight against climate hazards.

1.2 Conclusions

The presented theoretical and practical assumptions are to be used by the authors to achieve the research and analytical objectives. A wide range of research aims to comprehensively explain the impact of technological human ones, yet in the modern sense intelligent (Bostrom 2014). Neural networks operate in two interrelated stages. The first one is treated as learning based on a set of aggregated data describing the problem to be solved. This is done by means of appropriate learning methods that allow for the determination of the appropriate values of network parameters, i.e., the weights of connections between neurons and threshold values. The fundamental, core methods of learning multilayer perceptrons are the BackPropagation (BP) algorithm and the Quasi-Newton algorithm (variable BFGS metric). The fundamental element used to build a neural network is an artificial neuron, which is a model (albeit an imperfect and simplified one) of a biological neuron, which is an essential structural element of the brain. Neural networks in various applications have repeatedly proven to be an excellent prognostic tool (Tadeusiewicz et al. 2003, p. 109). Artificial neural networks generate a number of very good solutions for a multidimensional range of technical problems that can be defined (e.g., technical diagnostics of machines, devices and vehicles) They are used in elements such as the prediction and modeling of systems, where processes are and can be incomprehensible or very complicated. Artificial neural networks can be used to supervise processes that take place and are executed in real time, where input variables are direct readings of measurements used for online control, and the network learns control functions. Artificial neural networks also work well in specific problems that are too complicated for conventional technologies (e.g., problems that do not have an algorithmic solution, or if they do, it is too complicated to find), and they also work well where people are not able to apply known, traditional methods. Neural networks owe their exponential growth to two very important factors, namely internal, focused on the ability to generalize problems and predict information, and external, based on the unlimited availability of personal computers with enormous computing power. Table 3 below shows the types of artificial intelligence.

Table 3. Types of Artificial Intelligence.

No.	Types of Artificial Intelligence
1.	Artificial Intelligence
2.	Augmented Intelligence
3.	Safe AI
4.	Inverse AI
5.	Representational AI
6.	Predictive AI
7.	Narrow AI
8.	General AI
9.	Artificial superintelligence (ASI)

Source: Own study.

Table 4 below shows the different types of artificial intelligence categorized by learning technique.

Table 4. Types of artificial intelligence categorized by learning technique.

No.	Types of Artificial Intelligence – Learning Techniques
1.	Machine learning
2.	Deep learning

Source: Own study.

Table 5 below shows the types of AI models.

Table 5. Types of Artificial Intelligence Models.

No.	Types of Artificial Intelligence Models
1.	Deep neural networks
2.	AI logistic regression model
3.	Linear regression model
4.	AI decision tree model
5.	Random forest
6.	K nearest neighbor model (KNN)
7.	Naive Bayes Model

Source: Own study.

All types and kinds of artificial intelligence presented above can, if properly implemented, apply to creating positive climate change and building appropriate climate hazard management systems. The potential of different types and kinds of AI is huge and has a variety of mechanisms that can generate intelligent climate value for the climate.

2.4 Artificial intelligence and its application in modern technological and industrial systems

The economy, with its specific and peculiar conditions, is a dynamic system focused on multilateral relations. The economy is conditioned by numerous aspects of different nature and impact. The economy is constantly changing, adapting to the expectations of social groups, caused by their needs. The economy as a whole creates a coherent system of solutions and applications suitable for creating usefulness for the requirements of society. Skillful creation and stimulation of the economy is and should always be a priority for groups of decision-makers at political, business and social levels. These priorities develop the shape and image of technological and industrial systems determining

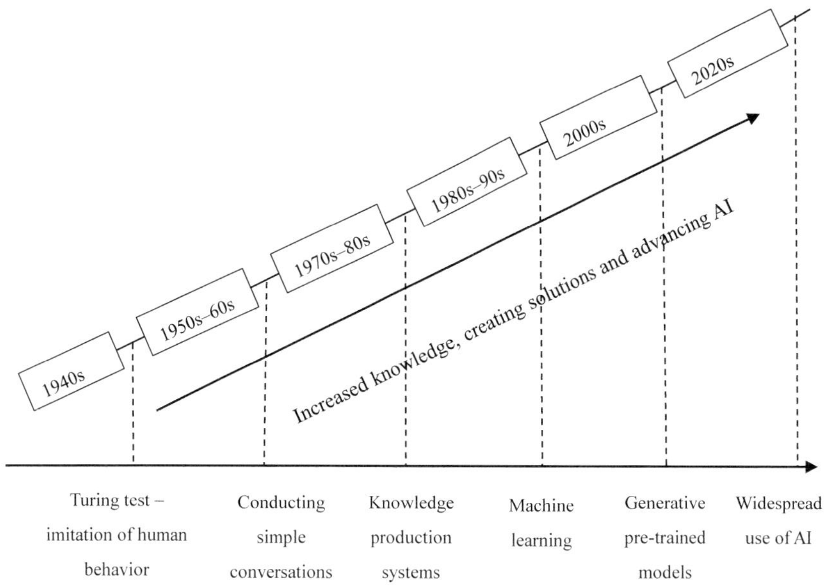

Figure 4. Stages of the development of artificial intelligence at the turn of the 1940s to the present day. Source: Own study.

2.6 Sources of the concept of artificial intelligence and its potential among climate-smart organizations

The development of artificial intelligence implies philosophical and existential problems, which are mainly ontological, epistemological and ethical (Orbik 2023). Every conceived concept always has its origin in the innovator's and visionary's mind (Buchanan 2005). In order to create something, first the vision of a creator of change and innovation is needed. It must arise in the mind and thought processes of the creator and then it must be materialized in the structures of technical and organizational capabilities. Such a path often requires experimental processes in a "from defeat to victory" system. This was the case for many discoveries on this earth, and it was probably also the case for concepts and later solutions related to artificial intelligence. The origins of the concept of artificial intelligence lie at the core of knowledge engineering, which is used here in a fairly broad sense in terms of the application of artificial intelligence, and concerns mainly the construction of so-called individual knowledge bases dedicated to a specific purpose. The primary task of knowledge engineering is to collect and formalize expert knowledge into the form of rules used by expert systems. Expert systems, as the first among numerous disciplines of artificial intelligence, have found wide practical application in business, the economy and society. Knowledge engineering and data mining are currently the two largest, core pillars of the design and construction of modern intelligent systems. Knowledge induction generated from data is very often based on the use of a wide range of machine learning algorithms and feature selection or extraction algorithms. The scope of artificial intelligence includes evolutionary algorithms, expert systems, heuristic, genetic algorithms, artificial neural networks and fuzzy logic. The system that connects the achievements of neural networks with expert systems is Knowledge-Based Artificial Neural Networks (KBANN). From this perspective, the potential of artificial intelligence can be aggregated for the development of so-called climate-smart organizations, which are entities that have built-in climate values in their DNA, i.e., the genotype of organizations. Climate values embedded in the genotype of organizations are the strategic helm shaping the structure of their business models and strategies, which directly translates into the logic of managing the processes and projects of these organizations aimed at satisfying climate criteria and principles.

The areas of interest of artificial intelligence are shown in Figure 5.

As Figure 5 shows, the development of artificial intelligence has a complex dimension based on broadly understood algorithmization supported by elements of heuristic, fuzzy logic and expert systems. In such a starry system, a multidimensional set of logically related technological solutions that generate the competency of artificial intelligence to create positive climate change is created.

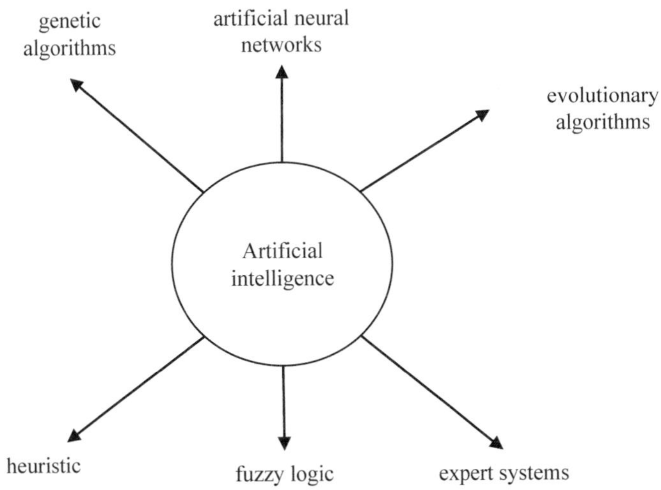

Figure 5. Areas of interest of artificial intelligence. Source: Own study.

2.7 The essence of artificial intelligence in the context of key definitions

Artificial intelligence, in its essence, is a technological concept, but it actually affects all factors of human functioning. Artificial intelligence can be interpreted, shaped, implemented and developed differently. The understanding and utilization of artificial intelligence can have many interpretations, meanings and notions. The question "Can a machine think?" can be referred to here. From this perspective, only a machine could think, and only very specific types of machines at that, namely brains and machines with intrinsic causal powers equivalent to brain powers. Therefore, strong artificial intelligence has little to tell us about thinking, because it does not concern machines but programs instead, and no program in itself is enough to think (Searle 1980). Artificial intelligence can be understood in social, sociological, legal, political and technical terms. Each of these interpretations generates a different meaning and designation. In the middle of this discussion is the applicability of artificial intelligence. A smart organization proposes a radically different way of designing an organization to harness the potential offered by modern information opportunities. A smart organization, designed backwards from their customers, considers information to be a key source of competitive advantage. A smart organization uses information to drive learning and adaptation at the level of individuals, tasks, processes, and the entire organization to deliver transformative changes in business performance (Beckford 2019). An organization's intelligence can be understood as raw material for meaningful signals and

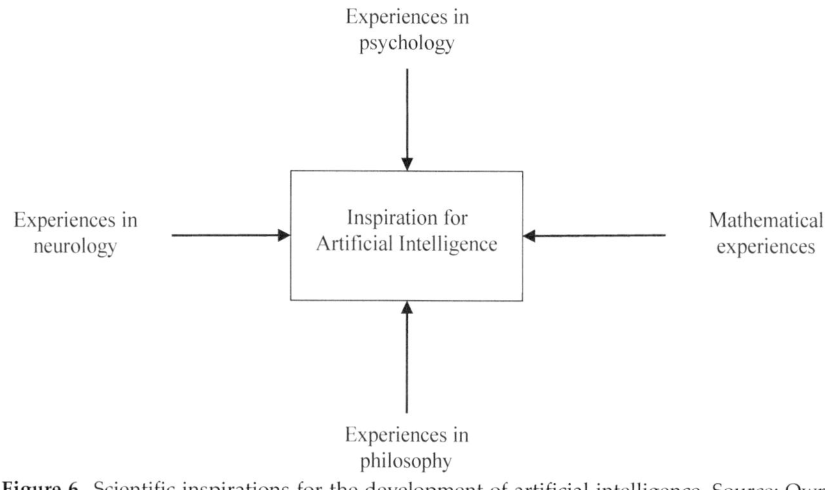

Figure 6. Scientific inspirations for the development of artificial intelligence. Source: Own study.

decision-making based on available content, various types of expertise mobilized in the organization, a set of processes aimed at acquiring and using intelligence throughout the organization and a strategic ability that allows the organization to operate effectively in a turbulent environment (Wilensky 1967). A smart organization can develop relevant skills from three perspectives: understanding the organizational environment, mobilizing resources appropriate to the environment and achieving goals (Matheson and Matheson 2001, pp. 49–54). Artificial intelligence has two core, crucial meanings. It is a hypothetical intelligence implemented in a technical process, rather than a natural one, and it is also the name of technology and the field of research into computer science and cognitive science, which also draws widely from the achievements of psychology, neurology, mathematics and philosophy, etc. Artificial intelligence is an imitation of real intelligence. Figure 6 shows the scientific inspiration for the development of artificial intelligence. These inspirations are important because they are focused, on the one hand, on experiences in neurology and psychology and, on the other hand, on the results of experiences in mathematics and philosophy. Such an interdisciplinary combination of intellectual capital inherent in each of these fields of science creates a complementary platform for the development of artificial intelligence.

Artificial intelligence can strongly support service design, which is a very important stage in the development of artificial intelligence solutions dedicated to business and the economy. The most important areas of service design focus on several design aspects, including the design of services, service processes and building the capacity of the service provider in such a way as to meet the needs and requirements of the recipient on the business side of the enterprise. In this area, the mechanisms of artificial intelligence work very well.

2.8 The issue of the consciousness of artificial intelligence in the processes of its use

When reflecting extensively on artificial intelligence, the following question becomes crucial: does artificial intelligence have consciousness? This is difficult because the concept of consciousness is difficult to define. Consciousness refers to sensory processes (vision, hearing, taste, touch and smell) and psychological processes (thoughts, emotions, desires and beliefs). In this context, it is necessary to remember that intelligence requires consciousness. So, the question is whether artificial intelligence will achieve an appropriate state of consciousness in its development process. An integrated information theory applies here, which indicates that any system, biological or artificial, which can integrate information to a high degree, can be considered conscious (Tononi 2004).

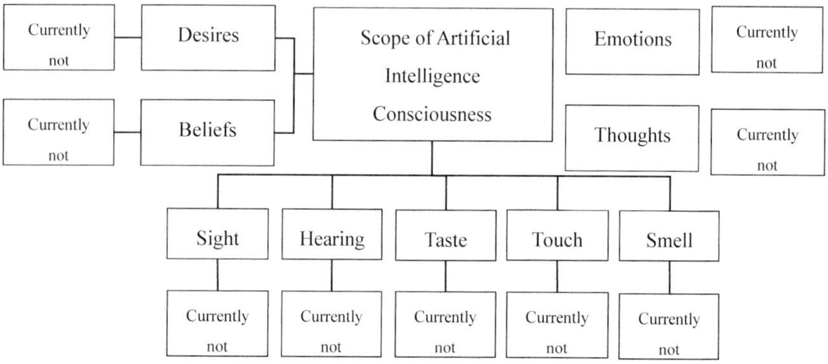

Figure 7. Future ranges of artificial intelligence consciousness. Source: Own study.

Figure 7 presents doubts about the identification of consciousness by artificial intelligence at the current stage of its technological development.

It must also be remembered that giving full consciousness to artificial intelligence can be and is very dangerous. First of all, it is related to the possibility of man potentially losing control of such artificial intelligence. This can generate a whole range of threats which are already increasingly visible, including to the existence of humanity itself.

2.9 Artificial intelligence in management sciences and climate sciences

Artificial intelligence is a very important field of scientific research at the intersection of computer science, neuroscience and psychology. Data science is an important part of the development of artificial intelligence. It involves in-depth analysis, interpretation, synthesis and the use of Big Data to draw appropriate conclusions and decision-making processes. Data science is an interdisciplinary field of research that uses scientific methods, processes and algorithms to extract value from data. It reveals trends and generates conclusions on the basis of which companies and other organizations can make more accurate decisions and create more innovative products and services (Adhikari et al. 2018). Artificial intelligence also uses sciences such as mathematics, statistics, psychology, philosophy, linguistics, neurobiology and robotics. Management sciences and climate sciences are two very important scientific disciplines that, from the perspective of climate change, can generate a great number of concrete and important theoretical and practical solutions for the space of human life. Management sciences and climate sciences are strongly intertwined due to dynamic anomalies and climate catastrophes. Climate sciences explain, among others, the effects of climate change, and management sciences give tools, concepts, methods and techniques for effectively and efficiently managing these changes and neutralizing their adverse effects. This is very important, especially in the current situation when different types of sciences create huge potential for interdisciplinarity. Therefore, the development of artificial intelligence draws its development possibilities from various sciences for the benefit of humanity. Artificial intelligence in the context of climate change develops and influences management sciences and climate sciences. Both of these scientific disciplines offer enormous intellectual potential for a proper understanding of climate change and its impact on numerous factors of human existence in the world. AI techniques make it possible to turn Big Data into useful insights and make predictions based on them (Nakamoto and Takasugi 2023). Figure 8 shows the use of artificial intelligence in climate change programs.

Artificial intelligence is strongly related to the concept of Big Data at the level of organizational and management mechanisms. Big Data is a concept that means vast, complex and diverse datasets which are constantly changing, and their storage and analysis is a complex process (Sagiroglu and Sinanc 2013). Artificial intelligence is also known as technological intelligence, and is transforming operations and decision-making in companies and public organizations across all sectors of the economy – and we are still witnessing the beginning of the AI-driven revolution (Brynjolfsson and McAfee 2014). Hence, artificial intelligence changes the rules of the game in management and quality sciences, especially in the area of creating strategies and business models, in relation to the new dimensions of building competitive advantage and competitive potential, including

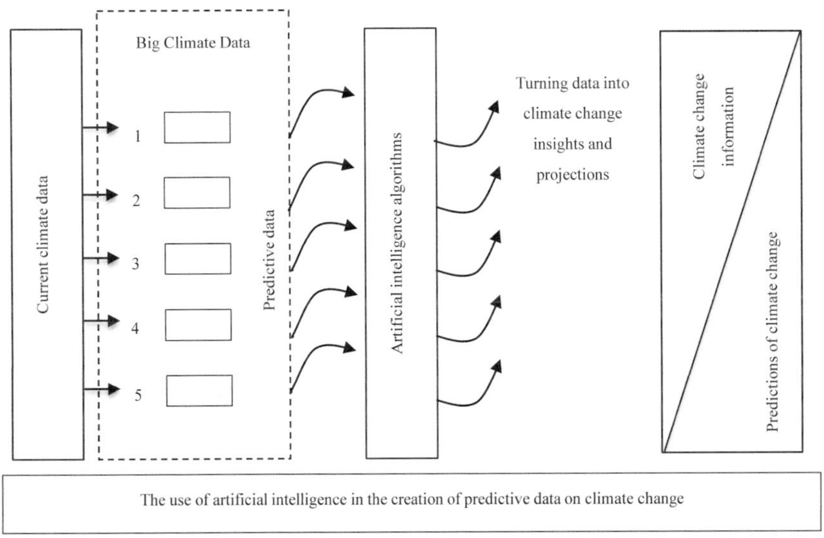

Figure 8. The use of artificial intelligence in climate change programming. Source: Own study.

technological advantage. It also covers the aspects of the various transformations of companies with the use of artificial intelligence. It marks a new impact of AI on finance, marketing, human resources, manufacturing and supply chains as new products and innovations develop. It also changes the rules for developing relationships with customers and other stakeholders.

2.10 Standardization of artificial intelligence in organizational and management processes

The standardization of artificial intelligence in organizational and management processes is of great theoretical and practical importance for the effectiveness and efficiency of this very important technological tool, with a specialized standard ISO/IEC 22989:2022 dedicated to it. This document establishes the terminology for artificial intelligence and describes AI concepts, and can be used to develop other standards and to support communication between different stakeholders. This document applies to all types of organizations (e.g., commercial enterprises, government agencies and non-profit organizations) (ISO/IEC 22989:2022). However, it should be noted that standardization is very difficult for highly developed and innovative solutions, which is related to the understanding of the concepts of innovation and quality. When an innovative solution moves into standardization processes, quality, as the sum of features and properties of the product or service striving to meet the anticipated and expected customer requirements, becomes a key word. Innovation is often imperfect and quality is repetitive. The transition phase in AI processes from innovation to standardized and repeatable quality can be very difficult or even impossible in the case of subsequent technological breakthroughs in these solutions. In the processes of applying artificial intelligence, it is very important to build a dedicated formalized management system. ISO/IEC 42001 is an international standard that specifies requirements for the establishment, implementation, maintenance and continuous improvement of an Artificial Intelligence Management System (AIMS) in organizations. It is intended for entities providing or using AI-based products or services, ensuring the responsible development and use of AI systems. The ISO/IEC 42001 standard provides organizations with the comprehensive guidance they need to use artificial intelligence responsibly and effectively, even when this technology evolves rapidly. Designed to cover different aspects of artificial intelligence and different applications that an organization can run, it provides an integrated approach to managing AI projects, from risk assessment to effectively addressing those risks (ISO/IEC42001:2023).

2.11 Artificial intelligence from the perspective of digital transformation as a new dimension of shaping the economic and social reality

The concept of artificial intelligence derives, among others, from the concept of cybernetics, i.e., the science of control, and from the concept of computer science, imposing its digital dimension strongly and clearly. Digital transformation, accelerated by the COVID-19 pandemic, among others, has unleashed a number of intellectual layers towards the development of the concept of artificial intelligence. In this sense, a completely new and surprising dimension of shaping economic and social reality is created. This dimension becomes a generator of an avalanche of changes and thus opportunities, benefits, and also threats, risks and challenges not necessarily in positive terms. This must be addressed primarily from intellectual and substantive perspectives. Only such an approach can unleash the enormous potential embedded in AI solutions.

What is important from the perspective of useful generated values is the search for consistency and scalability between these components of modern technology to achieve the maximization of benefits with minimal economic outlay and climate optimization because the dynamics of the processes in the economy, business and society create new needs in the dynamic development of contemporary organizations. This is especially true in a situation where companies, while existing in the so-called environment of many crises, are looking for new sources of competitive advantage. Modern companies must therefore, on the one hand, search for new core success factors of competitive advantage and, on the other hand, build a stable system of value creation in the short and long term. In order to be able to do this, it is necessary to properly assess the business reality in which the most important economic processes are currently conducted. Today's business situation, especially after the very difficult period of the COVID-19 pandemic, requires entrepreneurs to search for new technologies, new IT solutions, new digital marketing tools and, above all, a new way to allocate technologically and functionally innovative products and services expected by customers.

2.12 Conclusions

Considering artificial intelligence as an object of research in management sciences and climate sciences and as a practical tool for modern business solutions is fundamental to properly understanding its logic, methodology and its specially dedicated conceptual framework. This is particularly important for its applicability in effective and efficient management processes for shaping positive climate change. The proper targeting of the entire range of AI solutions towards actions in the area of eliminating and neutralizing climate change and building early response systems with the use of artificial intelligence seems to be a fundamental area of action and all related decision-making processes amid the deepening climate crisis.

References

Adhikari, A., DeNero, J., and Wagner, D. (2018). *Computational and Inferential Thinking: The Foundations of Data Science*, [on-line:] https://www.inferentialthinking.com.

Beckford, J. (2019). *The Intelligent Organisation (2nd ed.).* Taylor and Francis. Retrieved from https://www.perlego.com/book/1520133/the-intelligent-organisation-driving-systemic-change-with-information-pdf (Original work published 2019).

Bostrom, N. (2014). *Superintelligence: Paths, Dangers, Strategies*. Oxford: Oxford University Press.

Brynjolfsson, E., and McAfee, A. (2014). *The Second Machine Age: Work, Progress, and Prosperity in a Time of Brilliant Technologies*. New York, NY: W. W. Norton.

Brynjolfsson, E., and McAfee, A., (2017). *The Business of Artificial Intelligence What it can — and cannot — do for your organization.*, Harward Business Review, https://hbr.org/2017/07/the-business-of-artificial-intelligence.

Buchanan, B.G. (2005). *A (Very) Brief History of Artificial Intelligence*. AI Magazine 26, no. 4.

Campbell, J.A. (1986). *On Artificial Intelligence. Artificial Intelligence Review,* 1(1): 3–9. DOI: https://doi.org/10.1007/BF01988524.

Cichosz, P. (2000). *Systemy uczące się* [Learning systems], Wydawnictwa Naukowo-Techniczne, Warszawa, p. 40.

Communication from the Commission to the European Parliament, the European Council, the European Economic and Social Committee and the Committee of the Regions Artificial Intelligence for Europe {SWD(2018) 137 final, Brussels, 25.4.2018. COM(2018) 237 final.

Cornell, S., Berkhout, F., Willemijn, T., Tàbara, D.J., Jäger, J., Chabay, I. et al. (2013). *Opening up knowledge systems for better responses to global environmental change*. Environ Sci Pol. 28: 60–70.

Flach, P. (2012). *Machine Learning: The Art and Science of Algorithms.*

Gardner, H. (1983). *Frames of Mind: The Theory of Multiple Intelligences*. New York, NY: Basic Books.

ISO/IEC 22989:2022-Information technology — Artificial intelligence — Artificial intelligence concepts and terminology.

ISO/IEC 42001:2023, Information technology — Artificial intelligence — Management system Published (Edition 1, 2023).

Klein, M., and Methlie, L.B. (1992). *Expert Systems. A Decision Support Approach with Applications in Management and Finance*. Addison-Wesley Publishing Company.

LeCun, Y., and Bengio, Y. (1995). *Convolutional networks for images, speech, and time series, The handbook of brain theory and neural networks*, Volume 3361, Issue 10.

Lenton, T. M. (2021). *Tipping points in the climate system*. Weather. Vol. 76, No. 10, doi: 10.1002/wea.4058Earth Syst. Dynam., 15: 41–74, 2024 https://doi.org/10.5194/esd-15-41-2024.

Matheson D., Matheson J.E. (2001). *Smart Organizations Perform Better: Research Technology Management*, vol. 1, p. 49–54.

Matthews, J. B. R., Möller, V., van Diemen, R., Fuglestvedt, J. S., Masson-Delmott,e V., Méndez, C. et al. (2021). IPCC, Annex VII: Glossary, in Climate Change: The Physical Science Basis. Contribution of Working Group I to the Sixth Assessment Report of the Intergovernmental Panel on Climate Change, pp. 2215–2256.

Mulawka, J. (1997). *Systemy ekspertowe* [Expert systems], Warszawa, Wydawnictwo Naukowo – Techniczne.

Nakamoto, K., and Takasugi, K. (2023). *Special issue on application of artificial intelligence techniques in production engineering.* International Journal of Automation Technology, 17(2): 91–91.DOI:10.20965/ijat.2023.p0091.

Nilsson, N. J. (1998). *Artificial Intelligence: A New Synthesis.* San Francisco, CA: Morgan Kaufmann Publishers.

Orbik, Z. (2023). *W poszukiwaniu filozoficznych źródeł problematyki sztucznej inteligencji* [In Search of Philosophical Sources of Artificial Intelligence], „Ethos" 36, no 3(143) p. 45-66, DOI 10.12887/36-2023-3-143-05.

Sagiroglu, S., and Sinanc, D. (2013). *Big Data: A review, International Conference on Collaboration Technologies and Systems (CTS),* 42–47. DOI: 10.1109/CTS.2013.6567202.

Searle, J. R. (1980). *Minds, brains, and programs.* Behavioral and Brain Sciences 3 (3): 417–457.

Sprague, R.H., and Carlson, E.D. (1982). *Building Effective Decision Support Systems. Prentice Hall,* Englewood Cliffs, NJ.

Stocker, T. F., Jones, R. G., Hegglin, M. I., Lenton, T. M., Hegerl, G. C., Seneviratne, S. I. et al. (2024). *Reflecting on the Science of Climate Tipping Points to Inform and Assist Policy Making and Address the Risks they Pose to Society,* Surveys in Geophysics https://doi.org/10.1007/s10712-024-09844-w.

Tadeusiewicz, R., *Automatyzacja i sztuczna inteligencja jako źródła prawdziwych i wyimaginowanych zagrożeń,* [Automation and artificial intelligence as the sources of true and imaginary threats], Chapter in the book: Galwas B., Kozłowski P., Prandecki K. (ed.). (2019).: *Czy świat należy urządzić inaczej – schyłek i początek* [Should the world be organized differently – the end and the beginning]. Komitet Prognoz „Polska 2000 Plus" przy Prezydium PAN, Warszawa.

Tadeusiewicz, R., Izworski, A., and Stanisz-Walis, K. (2003). *Prediction of Clinical Diseases Using Artificial Neural Networks. In:* Prahlad Vadakkepat, Tan Woei Wan, Tan Kay Chen, Loh Ai Poh (Eds.). Proceedings of Second International Conference on Computational Intelligence, Robotics and Autonomous Systems CIRAS, Singapore, s. 109 (abstract), ISSN0219-6131, full paper (7 pages) on CD.

Tadeusiewicz, R., and Korohoda, P. (1997). *Komputerowa analiza i przetwarzanie obrazów* [Computer analysis and image processing], Wydawnictwo Fundacji Postępu Telekomunikacji, Kraków.

Tadeusiewicz, R. (1986). *Sygnał mowy i jego przetwarzanie oraz rozpoznawanie* [Speech signal and its processing and recognition], WT 9/86.

Tononi, G. (2004). *An information integration theory of consciousness,* BMC Neuroscience, 5: 42 doi:10.1186/1471-2202-5-42.

Wilensky, H. (1967). *Organizational Intelligence: Knowledge and Policy in Government and Industry.* Basic Books.

Chapter 3

Systematic Literature Review

Description of the Landscape of AI Research

3.1 Introduction

The purpose of this chapter is to present the results of a systematic literature review in relation to a description of the landscape of AI research used in climate hazard management.

The scope of the chapter covers issues related to the sources of artificial intelligence used in the management of climate hazards.

The subject of the chapter concerns a multidimensional cognitive perspective aimed at shaping the principles of artificial intelligence used in the management of climate hazards in relation to both theoretical cognition and implementation aspects.

3.2 Methodological assumptions for a systematic literature review

The study was conducted by means of a systematic literature review method to conduct scientific research into the dependencies of the AI research landscape used in climate hazard management. Scientific articles and gray literature were used. In order to be able to properly understand the principles of the use of artificial intelligence in the management of climate hazards, it is important to conduct a professional process related to the systematic literature review in order to obtain the appropriate characteristics of the landscape of artificial intelligence research.

DOI: 10.1201/9781003434016-3

In the course of the scientific discussion, two research hypotheses were formulated in this context:

Hypothesis 1: Artificial intelligence positively influences the identification of climate hazards affecting the formation of weather and climate anomalies.

Hypothesis 2: Artificial intelligence improves the identification of climate change affecting the emergence of multidimensional climate change.

Below is a description of the scope and subject of the research in terms of the verification of Hypothesis 1 – that artificial intelligence positively influences the identification of climate hazards affecting the formation of weather and climate anomalies.

In the relevant literature, the concepts of artificial intelligence, climate hazards and climate change are thoroughly characterized and described. The systematic literature review on artificial intelligence and the concepts of hazard and climate change aims to examine whether it is based on scientific evidence. For the purpose of conducting a systematic literature review, the databases were searched for the following terms:

- Artificial AND intelligence AND climate AND hazards. The Scopus and Web of Science databases were used. The following results were obtained: Scopus – 293 articles, WoS – 275 articles. A total of 105 duplicates were also generated.

Based on the research, a suitable Prisma model was developed for the terms 'artificial intelligence' and 'climate hazards', as shown in Figure 9.

For articles that were included, a semantic map of key concepts was prepared. The minimum number of occurrences of a keyword was set at five. Out of 1580 keywords, 67 met the threshold.

Table 9 provides a list of identified keywords related to the concepts of AI and hazards, taking into account the frequency of their occurrence in the analyzed scientific articles.

A detailed distribution of acquired concepts and their relationships is presented in the semantic map developed in VOSviewer, a software tool for constructing and visualizing bibliometric networks. These networks may include, for example, journals, researchers or individual publications, and may be constructed on the basis of citation, bibliographical coupling, co-citation or co-authorship relationships. VOSviewer also offers a text mining functionality that can be used to construct and visualize networks of co-occurrence of important terms extracted from scientific literature [https://www.vosviewer.com/].

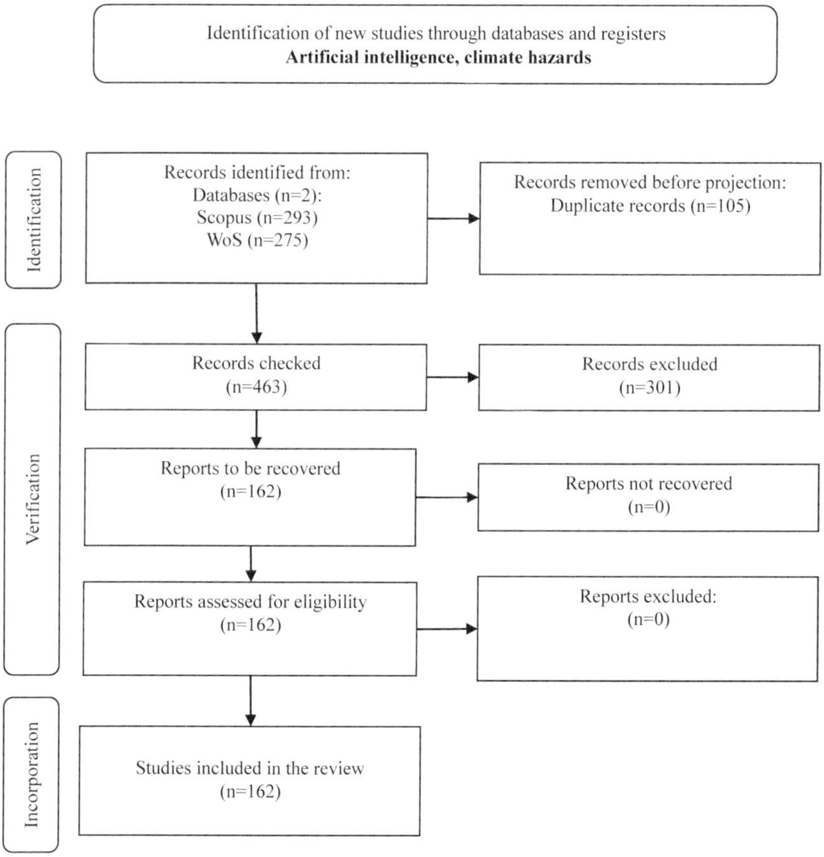

Figure 9. The Prisma model for the terms 'artificial intelligence' and 'climate hazards'
Source: Own study based on the results of the Rayyan program.

In the subsequent step, **Hypothesis 2 was verified: Artificial intelligence improves the identification of climate change affecting the emergence of multidimensional climate change.**

A further sequence of concepts was identified, namely: artificial AND intelligence AND climate AND change.

As many as 5044 articles were obtained from the Scopus database (this necessitated a narrowing of the scope to the Business, Management and Accounting categories, with 357 items obtained in this category**).**

4486 were obtained from the WoS database (this necessitated a narrowing of the scope to the Business, Management and Economics categories – 125 items were obtained in this category). There were 50 duplicates.

Based on the research, the Prisma model was developed for the terms 'artificial intelligence' and 'climate change'.

Table 9. List of identified keywords related to the concepts of AI and hazards, taking into account the frequency of their occurrence in the analyzed scientific articles.

Keyword	Occurrences	Total link strength
adversarial machine learning	5	33
air pollution	10	60
air quality	5	53
article	10	90
artificial intelligence	67	311
artificial intelligence (ai)	6	17
artificial neural network	7	49
big data	6	38
challenges	5	23
climate	11	40
climate change	48	266
climate-change	10	29
controlled study	5	42
decision making	8	59
decision support system	5	47
decision support systems	8	70
decision trees	6	59
deep learning	20	104
deforestation	9	56
disasters	10	57
drought	6	23
ecosystems	6	41
environmental monitoring	6	57
fire hazards	13	88
fires	8	58
flood	5	23
floods	5	39
forecasting	8	58
forest fires	6	44
geographic information systems	6	49
gis	7	41
global warming	5	23
hazard	7	50
hazards	9	51

Table 9 contd. ...

...Table 9 contd.

Keyword	Occurrences	Total link strength
human	5	45
impacts	8	26
learning systems	6	58
machine	5	14
machine learning	38	198
machine-learning	13	106
management	6	22
model	14	44
models	7	32
natural disasters	6	36
natural hazard	5	30
natural hazards	8	36
neural networks	5	47
neural-network	5	22
performance	5	28
pollution	5	27
precipitation	6	37
prediction	17	64
random forest	5	28
remote sensing	13	66
resilience	7	22
risk	8	30
risk assessment	17	125
risk management	5	27
simulation	5	24
spatial prediction	5	24
support vector machine	5	33
support vector machines	9	68
sustainable development	7	36
system	6	20
temperature	5	38
vulnerability	5	39
wind	5	40

Source: Own study

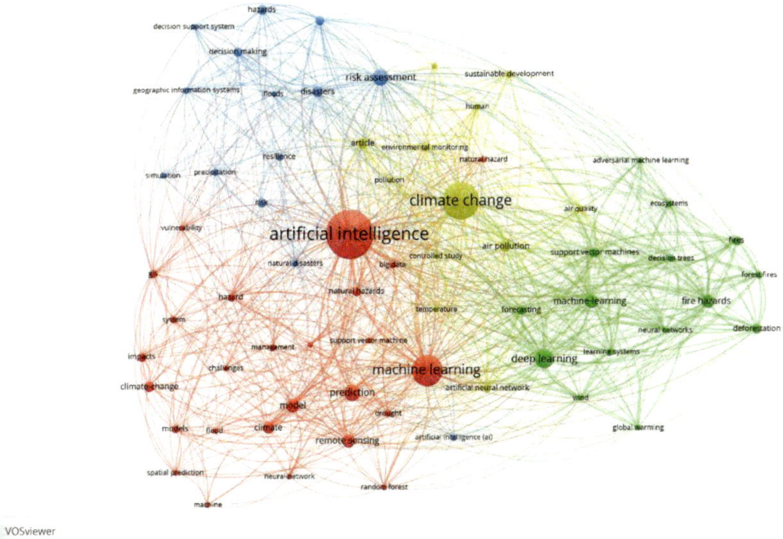

Figure 10. The semantic map of concepts related to concepts and their relationships is presented in the semantic map developed in VOSviewer. Source: Own study.

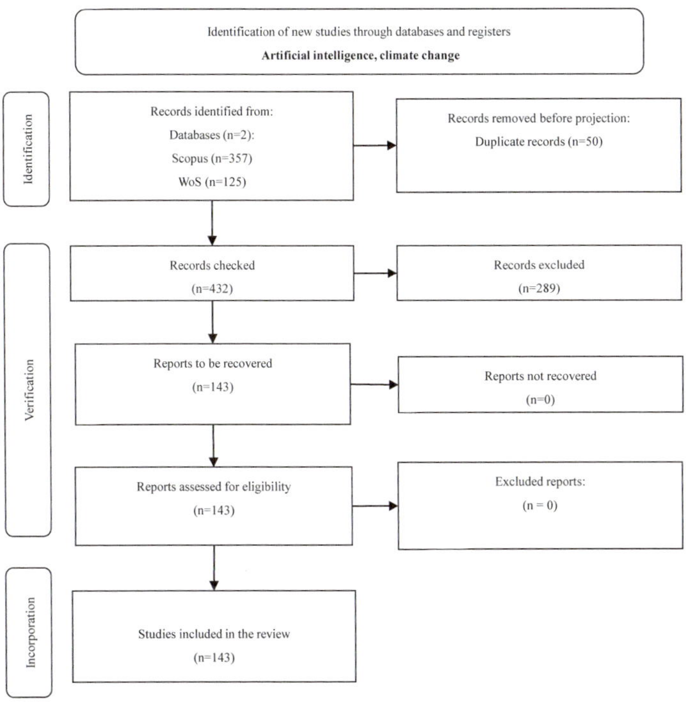

Figure 11. The Prisma model for the terms 'artificial intelligence' and 'climate change'. Source: Own study based on the results of the Rayyan program.

Keyword	Occurrences	Total Link Strength
adversarial machine learning	5	33
air pollution	10	60
air quality	5	53
article	10	90
artificial intelligence	67	311
artificial intelligence (ai)	6	17
artificial neural network	7	49
big data	6	38
challenges	5	23
climate	11	40
climate change	48	266
climate-change	10	29
controlled study	5	42
decision making	8	59
decision support system	5	47
decision support systems	8	70
decision trees	6	59
deep learning	20	104
deforestation	9	56
disasters	10	57
drought	6	23
ecosystems	6	41
environmental monitoring	6	57
fire hazards	13	88
fires	8	58
flood	5	23
floods	5	39
forecasting	8	58
forest fires	6	44
geographic information systems	6	49
gis	7	41
global warming	5	23
hazard	7	50
hazards	9	51
human	5	45
impacts	8	26
learning systems	6	58
machine	5	14
machine learning	38	198
machine-learning	13	106
management	6	22
model	14	44
models	7	32
natural disasters	6	36
natural hazard	5	30
natural hazards	8	36
neural networks	5	47
neural-network	5	22
performance	5	28
pollution	5	27
precipitation	6	37
prediction	17	64
random forest	5	28
remote sensing	13	66
resilience	7	22
risk	8	30
risk assessment	17	125
risk management	5	27
simulation	5	24
spatial prediction	5	24
support vector machine	5	33
support vector machines	9	68
sustainable development	7	36
system	6	20
temperature	5	38
vulnerability	5	39
wind	5	40

Figure 12. List of identified keywords related to the concepts of AI and climate change, taking into account the frequency of their occurrence in the analyzed scientific articles. Source: Own study.

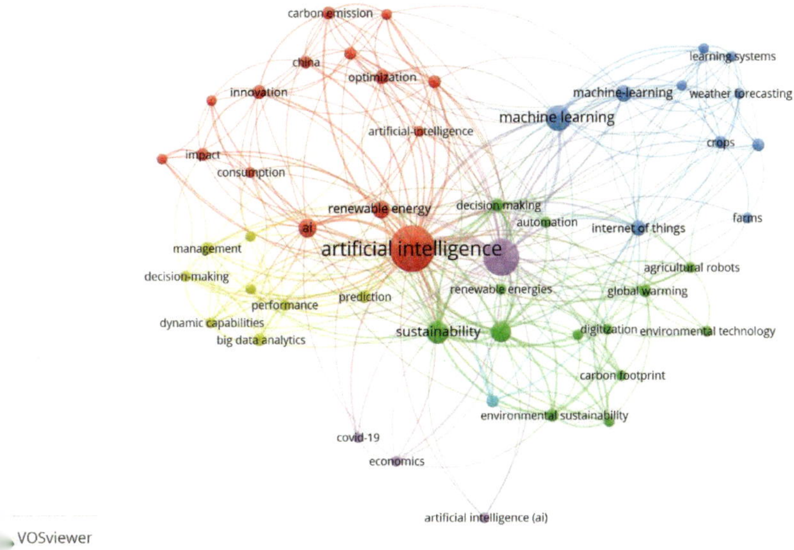

Figure 13. Semantic map of concepts related to artificial intelligence. Source: Own study.

For articles that were included, a semantic map of key words related to the subject was created.

A minimum number of occurrences of a keyword was used. Out of 776 keywords, 51 met the threshold.

Figure 13 shows a semantic map of concepts related to artificial intelligence.

The research based on the methodology of the systematic literature review (Bulgaria) has shown that both the relationships between artificial intelligence and hazards and between artificial intelligence and climate change are hot topics in the world literature. Both of these formulations generate a very large number of scientific publications related to numerous new ontological concepts which are important for the development of this subject. These include issues such as machine learning, optimization, sustainability, prediction, digitalization, carbon emission and dynamic capabilities, but also risk assessment, disasters, geographic information systems, natural hazards and others which are also important for the assessment of climate change predictions.

3.3 Summary of the systematic literature review and main conclusions

A systematic literature review of the theoretical and practical approach to the issue of artificial intelligence in managing climate hazards is fundamental to understanding its place and role in the business, economic and social existence of companies and other types of organizations and people in a strongly deepening climate crisis. Understanding the concept of artificial intelligence against the background of climate hazards at the level of global literature allows for a multi-criteria assessment of the impact of artificial intelligence on the activities of various types of entities for climate change and adaptation to it. Simultaneously, it allows for different opinions related to the benefits and costs of developing not only artificial intelligence in the economy, but also at the levels of individual countries, regions and cities.

3.4 Conclusions

The systematic literature review allowed for the determination of theoretical and practical assumptions for the process of climate hazard management. The proper selection of keywords and their importance open the way to understanding the very essence of the climate problem and extracting proposals for a strategic and tactical–operational decision-making process for it. Knowledge and science always create a special platform for proper substantive reasoning and correctly implemented inferences.

Chapter 4

Artificial Intelligence

Key Meanings and Interpretations in the Context of Climate Change

4.1 Introduction

The aim of this chapter is to present the key meanings and interpretations of artificial intelligence against the background of dynamic climate change.

The scope of the chapter covers the issues of artificial and non-artificial intelligence along with the mechanisms of applying Industry 4.0 and Industry 5.0 and the concept of Big Data. It also applies to bioinspiration, artificial and biological neural networks and evolutionary algorithms that can be used for effective and efficient climate management.

The chapter addresses a number of core issues related to artificial intelligence in dynamic climate management processes.

4.2 Artificial intelligence versus non-artificial (natural) intelligence from the perspective of climate change

Calling technological intelligence artificial intelligence is somewhat unfortunate – all the more so from the perspective of comparative analysis between artificial intelligence and human intelligence. Both types of intelligence are difficult to interpret. The same applies to the search for common and separate features and attributes between them. It is necessary to refer here to the concept of the Theory of Mind (ToM), a field of study that focuses on understanding and modeling the ability of the human mind to understand and predict other people's behaviors. This is an area that is essential for the development of artificial intelligence and has numerous challenges to overcome in the future. However, the mutual relationships between artificial intelligence and human intelligence

DOI: 10.1201/9781003434016-4

are strong and should not only be the subject of scientific research, but also the creation of these relationships and hybridization of intelligence as a joint development potential for business, the economy and society. Human intelligence allows for (i) the formulation and achievement of goals that help people achieve what they are looking for in life, taking into account their cultural context; (ii) making the most of their strengths and compensating for or correcting weaknesses; (iii) adapting, shaping and selecting environments; and (iv) the combination of the necessary skills (Sternberg 2003). Furthermore, it is linked to thought processes. From a hierarchical perspective, thinking has several layers: the first is the layer of form, which is language; the second is the content layer, which is the concept of consciousness. These two layers are an internal analysis of thinking itself. From the perspective of its interaction with the applied object, another layer of thinking occurs, i.e., its outer layer of functions. When these three layers are divided by functional structure, the concept of language and consciousness are internal structures, while the functional layer boasts the external function of thinking and consciousness, namely intelligence. The concept of consciousness, as the internal structure of thinking, refers to the structure of consciousness. Depending on the internal structure, reflective and control activities have corresponding manifestations at all internal levels of thought. At the content level, this is shown as activities of different types of consciousness, and at the form level as corresponding language activities (Dong et al. 2020). The American Psychological Association's (APA) report on intelligence says that "[i] ndividuals differ from one another in their ability to understand complex ideas, to adapt effectively to the environment, to learn from experience, to engage in various forms of reasoning, to overcome obstacles by taking thought" (Neisser et al. 1996). The types of human intelligence are shown in Table 10.

This is important mainly in connection with the taxonomy of people's learning systems.

The taxonomy of learning systems based on the amount of effort (inference) required from a learner to complete a learning task is as follows:

- learning by heart,
- learning from instructions,
- learning by analogy,
- explanation-based learning,
- learning from observations and discoveries.

Therefore, in such an approach, it is important to interpret the development of artificial intelligence against the background of intellectual capabilities possessed and used by people. Such comparative analysis should be carried out systematically and continuously through logically planned processes of measuring and monitoring differences in real time.

Table 10. Types of human intelligence.

No.	Types of human intelligence
1.	Logical–mathematical intelligence
2.	Linguistic (verbal) intelligence
3.	Interpersonal (social) intelligence
4.	Intrapersonal (intuitive) intelligence
5.	Interpersonal + Intrapersonal = Emotional
6.	Spatial intelligence
7.	Kinesthetic (motor) intelligence
8.	Naturalistic intelligence (otherwise natural or environmental)
9.	Musical intelligence (or auditory or rhythmic intelligence)
10.	Existential intelligence

Source: Own study.

4.3 Artificial intelligence against the background of climate change determinants

The phenomenon of artificial intelligence has many faces, images and interpretations. Artificial intelligence follows the principles of autonomy. The autonomy of AI systems should be understood as the ability to act independently without human intervention. It is a gradable feature, determined on the axis between its complete absence (the system is a tool in human hands and makes no decisions) and full autonomy (the system operates without any human intervention) (Rejmaniak 2021, pp. 95–113). Finding the right one is neither easy nor unambiguous, especially when it concerns the climate context, which includes a set of different requirements and expectations for halting adverse climate change. In this form, artificial intelligence gives rise to numerous opportunities and development perspectives for shaping appropriate adaptation strategies in the climate area, not only at the level of the state, regions and cities, but also in individual, wisely managed and climate-sensitive companies. Artificial intelligence functions in so-called cyber-physical systems, which are systems where the physical world connects with the virtual world through appropriate sensors and runtime modules and then processes information in the physical world by means of the mathematical mapping of physical objects. Figure 14 shows the relationships among cyber-physical systems in the context of AI principles. It should be noted that cyber-physical systems in the world of artificial intelligence are strongly connected with the mathematical mapping of physical objects and dynamic processing of information in the physical world through appropriate sensors and runtime modules.

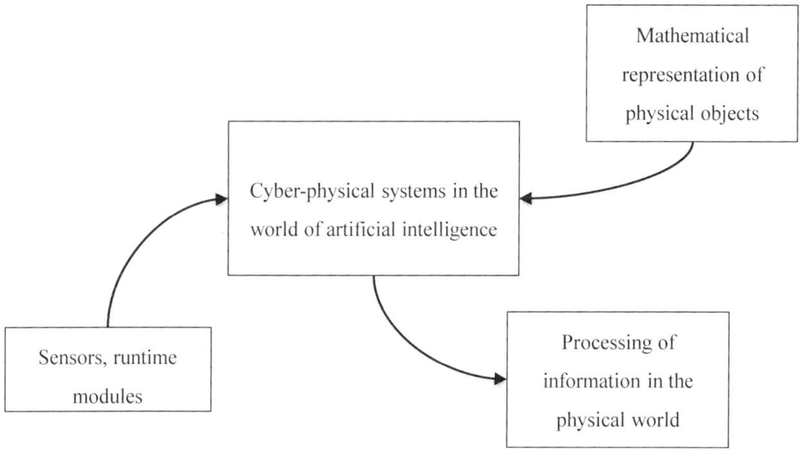

Figure 14. Relationships among cyber-physical systems in the context of artificial intelligence principles. Source: Own study.

In this approach, a number of interpretations related to artificial intelligence occur. Artificial intelligence is linked to the concept of cognitive computation, which is a technique that allows people to work with machines. The term comes from cognitive sciences and artificial intelligence. It builds algorithms with cognitive science theories to simulate human objective and psychological cognitive processes so as to enable machines to achieve a certain degree of "brain" cognitive intelligence (Chen 2017). Effective data management is a very important factor in AI processes, and this includes climate data as regards climate change. Fox and Hendler (2014) referred to the fact that the field of data science includes not only disciplinary foundations, but also strategies for solving real-world challenges. They gave details of four cross-sectional challenges in data science: understanding the scale in systems, distributed systems with incomplete and heterogeneous data, abductive reasoning and next-generation semantic data infrastructure (Fox and Hendler 2014). Artificial intelligence solutions use both deduction and induction. Deduction is a type of reasoning in which a conclusion is reached on the basis of premises. This means that we start with a set of true statements (or statements assumed to be true) and reach a new statement that necessarily follows from the premises. Deduction is mainly used in logic and mathematics. On the other hand, induction is a type of reasoning in which conclusions are drawn from the observation of a finite set of specific cases. This means that it starts by observing a limited number of cases and draws a general statement that covers them all. Induction is mainly used in science and philosophy. Therefore, from this perspective, artificial intelligence should strongly support and develop global climate ambitions at different levels of decision-making processes.

4.4 Industry 4.0, Industry 5.0 and artificial intelligence applications in climate management

Nowadays, the industrialization of the economy is undergoing significant changes in exponential terms. The dynamic development of the economy and its strong industrialization have generated a multitude of engineering applications combining technical and social systems, creating a highly interdependent socio-technical system, which is focused on two development areas. The first of these is focused on the dynamization of automation, robotics and technical diagnostics in technological production processes as part of the so-called Industry 4.0, while the second is focused on generating the connection of humans and their involvement in production processes with technological and social robots in order to enhance technological processes, their safety and cybersecurity and eliminate the human element from engineering difficult processes as the weakest link in socio-technical systems. **Industry 4.0** refers to the digitalization of industry, including the use of technologies such as the Internet of Things (IoT), automation, robotics, cloud computing and large-scale data analysis. Industry 4.0 is a combination of automation and data exchange in various technologies, the basis for which are cyber-physical systems using modern control systems with embedded software connected by IoT (Palwe and Sirsikar 2021). Cyber-physical systems (CPS) operate at the intersection of two digital and physical worlds. Cyber-physical systems are defined as computational processes operating in close interaction with a physical process (NSF 2006). Two related areas can be distinguished here, namely physical (physical, biological, etc.) and digital layers. Cyber-physical systems are intelligent systems that include designed, interoperable networks of physical and computing components. CPS and related systems (including the Internet of Things (IoT) and the Industrial Internet) are widely recognized as having high potential for innovative applications and impact in many sectors of the global economy (the National Institute of Standards and Technology). In the case of cyber-physical systems, the common denominator remains the ubiquitous ICT infrastructure. In the digital layer, calculations are made on the basis of measurements/observations of physical processes, while the results of these calculations affect the processes of the physical layer. The nine core key technologies that are pillars of Industry 4.0 include Big Data analytics, autonomous robots, simulation, horizontal and vertical system integration, the industrial Internet of Things (IIoT), cyber security, cloud computing, additive manufacturing and augmented reality (Para et al. 2018, pp. 121–134), (Reis and Geert 2017, Erboz 2017).

Industry 5.0 is an extension of Industry 4.0 and focuses on the human–machine interaction, aiming to create a working environment where humans and robots can work together safely and efficiently. Industry 5.0 is based, among others, on mechatronics solutions. In mechanical engineering, the use of artificial intelligence to automate routine tasks, optimize and streamline processes, and improve quality control and test precision is on the rise. The task of mechatronics is to produce optimal, multi-functional systems with a complex internal structure that operate intelligently but with uniform functional properties and systems in a changing environment. Therefore, solutions based on modelling processes and projects using artificial intelligence are becoming crucial. Artificial intelligence is also very much connected to cybernetics. It is based on the same theoretical foundations as cybernetics because it uses binary calculations and the Boolean function in its structure. To some extent, artificial intelligence is a kind of "cybernetics 2.0" extended with self-learning algorithms. The crucial difference between cybernetics and artificial intelligence lies in the approach to planning and purposefulness. In terms of cybernetics, it is crucial to first define a goal or destination, and then control the system so that this specific goal is fully achieved. Artificial intelligence has more freedom, but often operates in a more autonomous way, without a predetermined end goal, as is the case with cybernetics. Cybernetics generates key insights into the regulation of complex, adaptive social and technical systems. AI security is primarily about controlling certain behaviors in advanced, self-improvement algorithms. Cyber theories and concepts can provide clues about the mechanisms needed for this. Concepts such as circular causality and "second-order cybernetics" underline the role of the subject or observer in these systems. Cybernetics places a strong emphasis on how individual parts relate to a wider, designated system. In the case of artificial intelligence, it highlights and indicates a holistic, multidimensional dynamic approach to adaptation, rather than just applying censorship to generative capabilities or limiting specific behaviors. Cybernetic models balance defined stability and adaptation. Ethical AI requires the permanent introduction of dynamic principles and legal rules, which should responsibly evolve as technology advances. Cybernetics and artificial intelligence are therefore closely interlinked. The focus on cyber concepts provides numerous valuable perspectives for the responsible development of artificial intelligence in the future, especially with regard to Industry 5.0 concept solutions. Another important area in this respect is the relationship between robotics and artificial intelligence. Artificial intelligence in robotics is a combination of advanced algorithms and machine learning techniques with the capabilities of robots. Thanks to the use of AI, robots become capable of self-learning, adapting to new situations and making intelligent decisions. The connecting factor in this case is the rules of programming robots. Robot programming then involves creating software that controls the operation of machines, allowing them to communicate with other devices and process data. The core issues related to the programming of robots include the creation of control algorithms, which allows robots to perform tasks efficiently and precisely. Another component is defining communication interfaces, which is achieved by designing interfaces that allow robots to communicate with other devices, such as sensors or controllers. Safety is also important here, aimed at ensuring safety in the operation of robots, both for users and the machine itself. In the context of the development of artificial intelligence, an important component is the AI Agent analytical tool. Today's built-in functions allow for extensive autonomous activities in the area of automation of complex processes and the construction of modern relational systems. In addition, it should be noted that artificial intelligence mechanisms in production and logistics processes can support and solve overstock and out-of-stock problems. Proper adjustment of the internal needs of the company and customers with the help of artificial intelligence solutions fits comfortably into the new trends in the development of production and logistics systems.

4.5 Big Data and climate management

Multiple data sets are located in a dynamic web space. Managing multiple data sets is a great technological, organizational and management challenge. This is particularly important when creating an aggregated Data Management Center, grouping and processing data, which is unfortunately extremely energy-intensive. Big Data sets must be collected, verified and defined according to a certain position. Big Data technologies are changing our ways of thinking, working and living (Mayer-Schönberger and Cukier 2014). The concept of Big Data can be examined in terms of the 3V categories, namely volume, referring to the large volume of data that needs to be processed and stored; velocity, referring to the speed at which data is generated or the frequency at which it is supplied; and variety, referring to different data sources as well as their structured and unstructured nature (Wamba et al. 2015, pp. 234–246). Crucial challenges include the implementation of Big Data Behavioral Analytics (BDBA) to achieve agility, coordination, visibility, vigilance, customer insights and sustainability across operations and supply chains (Talwar et al. 2021, pp. 3509–3534). Understanding both the dark and bright sides of Big Data Analytics (BDA) in terms of climate change is crucial for informed decision-making facilitated by BDA and is key to a sustainable future (Papadopoulos and Balta 2022). Big Data technologies in some cases make it possible to analyze all data that relate to the phenomena studied, and not only, for example, a random sample, which is used in traditional quantitative research, assuming (trying to ensure) its representativeness. At the same time, the much greater detail and completeness of the data underpinning scientific cognition enables it to be highly scalable, from learning the general direction to learning the most specific details (Sosińska-Kalata 2018, pp. 7–35). Climate science, as a data-intensive subject, has been overwhelmingly affected by the Big Data era and the associated technological revolutions. The great successes of Big Data analytics in various areas over the last decade have also triggered expectations of Big Data and its effectiveness in addressing the larger problem of climate change. As a new topic, climate change has been at the forefront of Big Climate Data Analytics implementations, and extensive research has been conducted covering a variety of topics (Hassani et al. 2019). Table 11 below presents a list of selected climate database resources targeting Big Data.

The categorization and creation of Big Data can be seen as a response or counter-offensive to the emerging regulatory oversight of national governments and international organizations over the misuse of personal data ("human-created and human-related digital data") (Espinoza and Aronczyk 2021). The BDA has generally met the prerequisites and objectives of enhancing the adaptive capacity for climate change. Actual performance indicates that the BDA system works well in terms of processing speed, analysis reliability and data accuracy (Wang and Wang 2022).

Table 11. List of selected climate database resources focused on Big Data.

No.	List of selected database resources focused on Big Data
1.	Earth System Grid Federation (esgf.llnl.gov)
2.	NASA Global Climate Change (climate.nasa.gov)
3.	NASA Center for Climate Simulation (nccs.nasa.gov)
4.	Climate Data Guide (climatedataguide.ucar.edu)
5.	Global Climate Observing System (GCOS)
6.	National Center for Atmospheric Research (ncar.ucar.edu)
7.	United Nations Global Pulse (unglobalpulse.org)

Source: Own study.

4.6 Artificial intelligence in applications for co-production and climate change

Artificial intelligence in the processes of organizational, technical and operational changes can be a strong platform for accelerating efficiency and effectiveness at the level of various types of organizations. Co-production is one of the most well-developed components of climate management, generating many opportunities. One of the core analyses in the area of climate change is the analysis of how the replication process emerges from the co-production of technological solutions (TS) where AI artifacts are linked to companies and institutions (Lepratte and Yoguel 2024, pp. 535–558). Figure 15 shows artificial intelligence as an accelerator of change to improve the effectiveness and efficiency of organizations by using predictive factors of climate change monitoring.

With the help of AI-based predictive models applicable to co-production, it is possible to predict future climate change with greater accuracy, which is invaluable in planning actions to limit the effects of global warming. Artificial intelligence also helps to identify and classify waste on a much larger scale than ever before, which significantly improves the entire recovery process of raw materials. Moreover, smart waste management systems can predict waste generation in different areas. The use of intelligent algorithms to analyze trends in pollution resulting from production processes also allows for more effective planning of technological processes. Predictive business process monitoring is designed to predict the results of current business process cases based on past track records. The literature proposes a wide range of techniques for this predictive task. It turns out that no single technique, with the default configuration, consistently achieves the best predictive accuracy in all

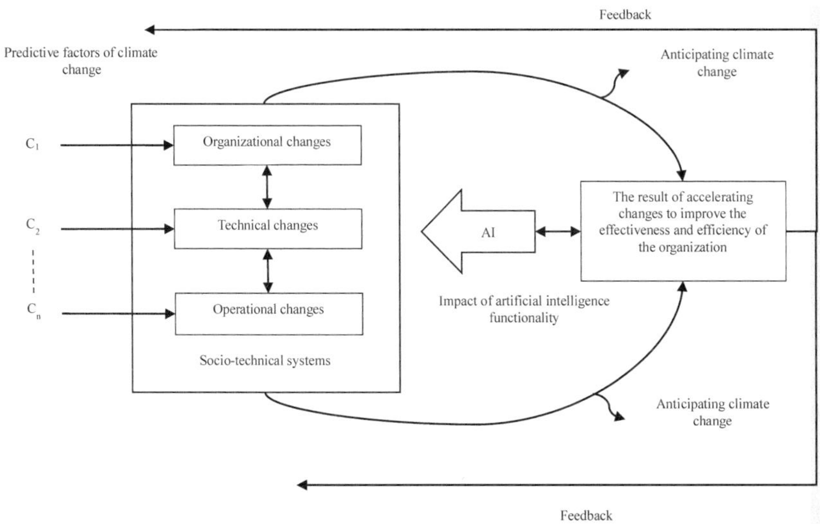

Figure 15. Artificial intelligence as an accelerator of change to improve the effectiveness and efficiency of organizations through the use of predictive factors of climate change monitoring. Source: Own study.

data sets. Therefore, the choice and configuration of the technique must be made for each data set (Di Francescomarino et al. 2018, pp. 67–83).

4.7 Superintelligence based on synthetic data and its use in climate management

In futuristic terms, the concept of superintelligence has become the subject of numerous approaches, analyses and applications. Visionary creativity has repeatedly determined the possibilities of hyper-intelligence as something extraordinary and surprising and at the same time very exciting. Superintelligence is any intellect that far surpasses human cognitive performance in virtually every field (Bostrom 2003). The way to superintelligence is, among others, brain emulation, which involves the creation of intelligent software by scanning and modeling the computational structure of the brain. Another way is to strengthen biological brains, and the best way to do so is gene manipulation. The next proposal is a brain–computer interface, which may potentially take the form of implants. A potential future pathway is the gradual improvement of networks and organizations connecting individual human minds through various methods, such as automating programs (Chalmers 2010). In this important context, Stanisław Lem (Lem 2020) draws attention to the concept of intellect, cerebromatics and intelligence amplification. Superintelligence can be based on synthetic data, which is computer-generated data that is quickly becoming an alternative to real-world data. Computer algorithms are not collected from actual documentation, but generate synthetic data. Synthetic data is artificially generated by means of algorithms or computer simulations that statistically or mathematically reflect real-world data. Figure 16 shows the use of superintelligence in climate change monitoring and prediction.

According to this interpretation and the adopted scientific reasoning, the use of superintelligence for monitoring and predicting climate change is supported by multidimensional computer simulations with reference to synthetic data. The whole process uses computer algorithms that generate synthetic data.

4.8 Bio-inspiration and self-organizing systems in artificial intelligence and climate management

Biology, nature and geography are sciences that are strongly correlated and provide powerful potential to strengthen strategic visions of human development and existence on earth. All of this opens up space for spatial and abstract thinking to create and materialize a vision of creation that draws ideas from the natural ecosystem. The bridge between the world of plants and animals and technology is created by a branch of science called bionics. The name comes from the Greek words "bios", meaning life, and "mimesis", meaning imitation. The applications of bionics in the economy and society are extremely inspiring and diverse and concern almost all areas of human existence. Bionics is based on thinking by analogy through the search for analogous phenomena and through the use of natural analogy to solve specifically defined problems. Bionic technologies are evolving rapidly. The biotechnology industry is growing at an impressive pace, and the pace of growth determines the availability of new technologies. Modern biotechnology, with the help of modern technologies and methods, allows for the introduction of modifications in living organisms so that they become more useful in our lives. Translating the functioning of the human brain into artificial intelligence is based on the concept of mapping and analogy. It is necessary to be aware that the brain is a plastic and dynamically changing creature that dynamically reacts to complex situations. It is capable of constant adaptation and self-organization. It also has the ability to compensate for damage to components and is capable of processing multiple data in parallel.

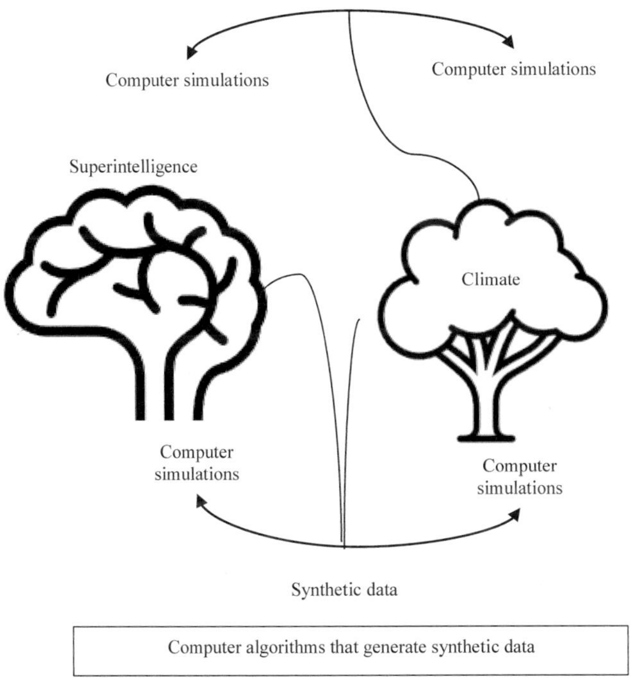

Figure 16. Superintelligence in the process of climate change monitoring and prediction. Source: Own study.

The brain also enables learning and generalizes objects observed with various stimuli, has full adaptability, distinguishes emerging acts and rules and generates thinking processes. Moreover, it reacts quickly and creatively in emerging new contexts. Thanks to this, it constantly and actively represents knowledge therein and is a comprehensive habitat of existing intelligence. Real biological neurons form a network in the brain with a very complex structure, the tracing of which required many years of efforts by numerous researchers, and this process is by no means yet definitively completed. However, for computational purposes, a network where artificial neurons are connected according to certain arbitrary rules, guaranteeing simple (and therefore cheap) implementation of artificial neural systems and high efficiency in solving practical tasks ordered to such a network is sufficient (Tadeusiewicz 1993). Bionics also deals with the modeling of human biological systems and processes. Bionics is biologically inspired engineering, as well as an application of biological methods and systems for the study and design of technical systems and modern technologies. It is used to develop technical solutions that function as living organisms or parts of these organisms. As part of the adopted solutions using bionics, the concepts of biomorphism, biomimicry and biomimetics occur (Benyus 2003). Biomorphism is treated as the definition and description of certain views related to the philosophy of nature by accurately grasping the mental, social and sometimes even physical phenomena as specific forms of life. Biomorphism can also be understood as the search for and development of created solutions similar to those found in nature without real imitation of the natural form, shape, process or system. Biomimicry is regarded as the imitation of appropriate models, systems and elements of nature in order to solve even the most complex human problems in such a way that does not in any way disturb the functioning ecological balance. On the other hand, biomimicry is focused on the proper imitation of effective, natural solutions. This is a crucial concept, which consists of seeking constant inspiration and solutions to problems occurring in the nature around us. It involves the diverse observation and study of natural processes occurring in nature as well as systems and structures in such a way as to understand how to design and create, among others, sustainable technologies and innovations that can bring far-reaching benefits to society and the environment. The term "biomimicry" is derived from the Greek words "bios", meaning life, and "mimesis", meaning imitation. The trend of scientific thought so defined is primarily aimed at directly drawing inspiration and a logical understanding of nature in order to solve the problems that man is currently facing, but which nature has already solved. From the development of advanced technology to simpler systems, biomimicry is focused on learning from proven nature strategies and putting them into practice. Artificial intelligence can directly benefit from such an approach and inspiration.

The most significant core benefits of biomimicry include prevention of excess resources and optimization of systems. As a result, nature-inspired solutions help avoid the overexploitation of existing resources. In addition, energy and resources are used more efficiently, mimicking natural processes that are inherently efficient. In addition, better adaptation to the environment generates solutions that are more flexible and adaptive based on nature, which means that they are better able to cope with changing situations and various types of phenomena. Biomimetics, on the other hand, is an interdisciplinary creative process of solving human problems through the abstraction, transfer and application of knowledge based on biological models. The term biomimetics was coined in the 1950s by the American inventor, biophysicist and polyhistorian Otto Schmitt. Biomimetics is the extraction of good projects from nature. It comprises the way from biology to engineering. Strategies based on natural patterns can be searched for by means of the Theory of Inventive Problem Solving (TRIZ), which requires certain skills and experience to use. It is not a machine that accepts problems at one end and spits out solutions at the other. In this respect, a computer program is probably not sufficient for a truly efficient use of the system (Vincent and Mann 2002, pp. 159–173). The search for natural and artificial patterns can be carried out using software based on the SAPPhIRE model in the form of VNA (verb–noun–adjective) and/or functional modeling. The following can then be observed:

(a) Conceptual design is an important stage of design.
(b) Laws and physical effects are important in design and can be used at the conceptual stage, but are currently not used properly.
(c) The SAPPhIRE causality model uses laws and effects, but has not been studied in detail for analysis and synthesis.
(d) The lack of clarity in the current definition of model design can make understanding the model difficult.

Therefore, the objectives are as follows:

(i) Explaining the design of the SAPPhIRE model.
(ii) Exploring the capabilities of the model for handling analysis and synthesis (Srinivasan and Chakrabarti 2009).

All these proposed approaches constitute an appropriate process of bio-inspiration, which is effectively and efficiently used in the development of the concept of artificial intelligence.

4.9 Artificial and biological neural networks as a tool for self-organization and the creation of structures capable of learning, adaptation and creativity and climate management

The network is a specific dependency system with a nodal link structure. It is examined, among others, in biological and technical terms. The network and its nodes generate a network effect relevant to their own capabilities.

The term "neural networks" refers to complex structures of the nervous system that work together to process information and perform complex activities. They consist of billions of neurons connected by synapses and forming an extensive system of connections. The biological basis of neural networks is a fascinating combination of biology and technology that inspires the development of artificial intelligence. Neurons, the basic nerve cells in the human brain, play a crucial role in processing information. Each neuron consists of a cell body, dendrites, and an axon. Neurons communicate with each other by means of electrical and chemical signals that flow between them through synapses, which are microscopic connections that allow signals to be transmitted from one neuron to another. When the signal reaches the end of the axon, it releases neurotransmitters, or chemical molecules that cross the synapse and stimulate or inhibit another neuron. Through these processes, the human brain is able to perform complex tasks such as thinking, feeling emotions and making decisions. The application of artificial neural networks as expert systems should most often be sought in such areas of the economy that are quite poorly formalized, i.e., it is difficult to describe them in a clear mathematical way and apply a strict algorithm of operation for them, and where the time of building the system is important. Learning artificial neural networks often replaces the knowledge of formal relationships between processed data. Heuristic methods are among the basic tools of artificial intelligence. In recent years, a significant number of methods have been designed and developed that belong to the group of the so-called metaheuristics, which are mainly inspired by biological and physical mechanisms. The most well-known methods include evolutionary algorithms, including genetic algorithms, the ant *colony optimization* algorithm, the taboo search algorithm and particle swarm optimization.

Evolutionary algorithms (EA) include specific methods that are inspired by the principle of natural selection, in which appropriate selection, reproduction and mutation mechanisms are used, as is the case in the biological evolutionary process.

Genetic Algorithms (GA) are the best known and most widespread class of evolutionary algorithms. Genetic algorithms search the space for alternative solutions to the problem in order to find the best possible solutions on the basis of specifically applied modelling of mechanisms of biological evolution (Goldberg 2003, Michalewicz 2003). Genetic algorithms are metaheuristic optimization algorithms that resemble natural evolution. Based on evolutionary theory of survival of the fittest and ideas of selection and mutation, genetic algorithms aim to simulate the evolution of solutions over different generations to ultimately identify the optimal or near-optimal solution to the optimization problem (Mitchell 1995, pp. 31–39).

The Ant Colony Optimization (ACO) algorithm is a technique dedicated to solving designated problems solved by searching for good ways in graphs, inspired by the behavior of ants in nature looking for food for their colony. In ACO algorithms, a colony of artificial individuals cooperates continuously in the search for optimal solutions (Dorigo and Stützle 2004).

The Taboo Search Algorithm (TS) is a dedicated metaheuristic that seeks to minimize the function of a goal. This algorithm is based on the search for random spaces of specific solutions, but it also has the so-called "taboo list", i.e., prohibited solutions, which importantly enables avoidance of being stuck in the local minimum and directs searches to other more favorable areas of space (Glover 1997, pp. 1–75).

Particle Swarm Optimization (PSO), also known as swarm intelligence or distributed intelligence, is a specific probabilistic optimization technique inspired by the behavior of a herd or flock of animals (insects, birds, fish). Swarm intelligence is a concept that means the formation of cooperation between many natural participants (e.g., a flock of geese) or artificial participants (e.g., a group of robots) without a predetermined, defined plan and without specifying the commanding unit. This phenomenon consists primarily of collective self-organization and is precisely the basis of metaheuristics (Eberhart et al. 2001).

To sum up, nature-inspired heuristic algorithms concern group/swarm intelligence – ant *colony optimization* algorithms, multimolecular optimization and immune systems – natural and artificial immune systems, the clonal selection algorithm, the negative selection algorithm and differential evolution.

4.10 Evolutionary algorithms in climate change management processes

An algorithm is an aggregated and finite sequence of clearly defined actions necessary to perform a certain type of task, which is also a way of dealing with a specific problem. In the construction and implementation of algorithms, the assumptions of algorithmics are used. Algorithmics is a specific unit of computer science dealing with the theory related to the creation and study of algorithms, correctness, complexity and algorithmization of processes and phenomena in the world around us. The creation of an algorithm can be defined as a specific process of modeling a certain phenomenon or process for a certain type of input to be properly processed in order to achieve a certain result in a finite and quantifiable number of steps. Aggregated algorithmization solutions are used here. Algorithmization is the process of building an algorithm for a selected, specific phenomenon or process that we want to model and automate, for the purpose of its full implementation and execution on a dedicated computing machine. So-called evolutionary algorithms are used in the processes of solving various types of problems with the use of artificial intelligence.

Evolutionary algorithms (EA) are a search and optimization technique based on principles taken from the theory of evolution. Their naturalness and simplicity of operation have made them highly popular in management and quality sciences, used to solve problems of combinatorial optimization, and in particular broadly understood problems of resource allocation (Różanowski 2007, pp. 109–135). The basic types of evolutionary algorithms are the simple genetic algorithm (SGA), evolutionary strategies and evolutionary programming. Evolutionary models such as Darwinian evolution, evolution as an inspiration for solving optimization problems, biological terminology and the fundamental concepts of evolutionary calculations apply here. The general scheme of the evolutionary method includes initialization, assessment of adaptation, selection of individuals, genetic operations and conditions of detention. Encoding methods in the processes of application of evolutionary algorithms include binary, logarithmic, real-valued, integer and permutation coding. The methods of selection used are proportional selection, rank-based selection, threshold selection, tournament selection and elitist selection. Recombination encompasses single-point and multi-point crossover, uniform crossover, arithmetic recombination and averaging. Mutation is applied to binary and real-valued encoding. The dynamics of local and global adaptation involve the adaptive landscape of adaptation functions, typical landscapes, selective pressure, local and global dynamics, and exploration and exploitation.

4.11 Real-time self-organization and adaptation mechanisms in climate management processes

Regulation in classical terms is understood as maintaining something in a correct and accepted state. Regulation can be automatic and can then be called self-regulation, which is the ability of a system to endogenously ensure its *equifinality,* i.e., that the system reaches a predetermined end state regardless of the current state, the course of internal processes and disturbances arising from the environment. Self-control is a stronger property of the system that allows it to determine this endogenous state. Self-regulation refers to conscious control of behavior that surpasses only strong will and the mere pursuit of goals (Gollwitzer and Moskowitz 1996, pp. 361–399). Self-regulation can be stable and dysregulated. After all, it can be full or partial self-regulation. Self-regulation can also be combined with co-regulation. Self-regulation is important in decision-making processes and plays a strong role in technical systems. In the systems concept, self-regulation, lack of self-regulation or positive and negative self-regulation can occur. Self-regulation is linked to the concept of synergism (from the Greek συνεϱγός – acting simultaneously), which underlies management. It is borrowed from synergetics, the theory of self-organization of complex dynamic systems, which is a philosophical stage in the development of science. The concept of synergism envisages the joint operation of a number of elements, which results in the agreed mutual activity of these elements resulting from disorderly states. Synergetics can be the basis for making effective management decisions in conditions of instability of the operating environment of various types of systems.

4.12 Irresponsible artificial intelligence and its risks in effective climate management

Artificial intelligence has a huge impact power, and unfortunately also an adverse one. Everything takes place at the level of intentions using this tool. If the intentions are positive, the effect of the applications will be positive as well.

However, if the AI user has bad intentions, the consequences of his actions can be very dangerous in many dimensions. Therefore, a vitally important area of development of artificial intelligence is the sanctioning of its use at the regulated and legal levels, where these two are linked by the ethical level. The formal and legal system, on the one hand, should protect humanity from negative actions by means of artificial intelligence and, on the other hand, should not kill its flexibility and capability of agile creation. Therefore, it must be borne in mind that growing cyber threats and the perception of emerging cyber uncertainties are important from the perspective of possible so-called irresponsible artificial intelligence.

The AI system can be used as a crime tool – crimes with AI – AI as a tool. It is becoming one of the favored modi operandi of perpetrators of crimes not only related to cyberspace; they can potentially change and develop ways of committing hitherto known crimes, or also create new threats to the security of the state, society or citizens. Crimes on AI – AI as attack surface occur when AI systems can also be the executive object of an action. Knowing the vulnerabilities and limitations of the algorithms used in the system, its proper functioning can be influenced and the following assumed criminal goals can be achieved: from slowing down, deceiving, blocking the system, to impersonating its administrator and taking control of the system or executing a criminal's "command", e.g., embedded in "poisoned data". Crimes by AI – AI as intermediary – This category is the most controversial, as it addresses the issue of the possibility of attributing legal personality to AI systems, which may also involve liability (Hayward and Maas 2021, pp. 209–233).

In military terms, robotics and artificial intelligence in armed conflicts pose new challenges for building and strengthening peace between nations and for preventing wars and armed conflicts in general. New rules and political and legal arrangements are needed, but they are developing too slowly (von Braun et al. 2021, pp. 1–13).

In such a narrative, it is worth asking the question: will artificial intelligence save or destroy the climate? This is a core problem of artificial intelligence. For example, the computing infrastructure built for the AI tools developed by companies entails a sharp increase in energy demand and, consequently, greenhouse gas emissions associated with the company's operations. In this case, it is important to design more efficient AI algorithms and dedicated learning methods, including the search for suitable AI alternatives to machine learning, which usually requires processing a very large amount of data.

4.13 Ethics and the legal dimension in the world of artificial intelligence as a component of digital and technological security in socio-economic ecosystems focused on climate change

Artificial intelligence ethics means human ethics towards artificial intelligence. The word ethics comes from the Greek word ethike and means morality (ethikos is moral, ethical, from ethos – a custom). In addition to physics and logic, Aristotle introduced ethics into ancient philosophy (384–321 B.C.) – he called it practical philosophy. Artificial intelligence solutions should be examined from an ethical perspective. AI solutions can be dangerous in two streams. The first stream results from the developed technological solutions and the second stream depends on the degree of use of artificial intelligence in practice. It all depends on the ethical values of the people and entities that can have a real impact on these two streams. Ethics in the context of artificial intelligence covers a wide range of issues that relate to morality, integrity, accountability, transparency and fairness in the creation, implementation and use of AI systems. The morality of algorithms and robots in AI solutions is of fundamental legal and philosophical importance. The morality of algorithms lies in how these algorithms make decisions and whether their results comply with the ethical standards of society. Ethics in artificial intelligence also requires reflection on data collection, storage and processing. Algorithms can generate, among others, social inequalities. Algorithms based on so-called training data can reproduce and strengthen current social inequalities. The transparency of the algorithms being built is not entirely correct. They often turn out to be black boxes. A great number of advanced AI algorithms act as black boxes, which means that the decision-making process is not understandable to the ordinary user in many cases. The issue of liability for possible errors arises. How should liability for errors made by AI systems be attributed? Should this be a task dedicated to programmers, businesses, or maybe states and regulators? Self-correcting algorithms are currently being developed. How can self-correcting mechanisms be implemented so that algorithms can learn on an ongoing basis and adapt to changing conditions, thus minimizing the risk of making mistakes in them? The same applies to robots. We are dealing here with the ethics of machines (or the morality of machines). This is a field of research dealing with the design of artificial moral agents (AMA), robots or computer programs that behave morally or immorally. In this context, the term roboethics is also used. One of the core assumptions of robotics is to determine the method and degree of optimal human control of robots. According to Gianmarco Verruggio (Pransky, pp. 6–10), (Veruggio 2005), these are two closely interlinked priorities. Ethics for scientists constructing robots must be developed as well as artificial ethics that will be programmed in robots. Researchers need to start analyzing these kinds of issues and examine whether regulations or laws are needed to protect citizens. Robots will become more intelligent and, in some ways, more powerful than human intelligence. However, it will still remain alien intelligence. Therefore, priority should be given to people (Habershon and Woods 2006). Table 12 below lists the core laws of robotics.

In the context of defining the laws of robotics, D. Jemielniak defines the fourth Law of Robotics, which, according to him, reads: 4. A robot must not mislead a person by impersonating another person (Jemielniak 2025, p. 16).

The laws of robotics can also be extrapolated to cyborgs. What is a cyborg? The word itself, which is a portmanteau of the words cybernetic and organism, was coined in 1960 by Manfred Clynes and Nathan Kline and means a being with biological and technical elements, usually a man connected to a machine (https://en.wikipedia.org/wiki/Cyborg). What issues should be analyzed in the first place in terms of the future law of cyborgs? Undoubtedly, the issue of data protection comes to the fore. The second fundamental issue of cyborg law is human autonomy. When faced with machines, separating what the so-called free will of man is from what the will of the machine – artificial intelligence – is, which tells man what to do, will be a fundamental challenge for the law (Księżak 2021). The core aspects of cyborg law are also addressed by Woodrow Barfield and Alexander Williams (Barfield and Williams 2017, p. 15), (Barfield and Williams 2017, www.mdpi.com/journal/philosophies). It should be remembered, however, that humanity and its essence are fundamental in the development processes of artificial intelligence (Tegmark 2017). Another important area of regulation in this area should concern bots. Solutions are needed to determine the identification and traceability associated with bots. Identification, in this sense, should be defined as conferring a permanent feature on bots, and traceability should ensure full traceability of the bot's functioning history. The idea is also that a bot or robot, through its improper programming or lack of appropriate safeguards in these processes, can mislead a person, e.g., by *de facto* impersonating a real person. This phenomenon may occur if we build so-called autonomous robots and/or bots. Security systems should be built into the appropriate software, which should also be fully identifiable and traceable with the complete possibility of assigning this software to a given robot and/or bot.

Table 12. List of the core laws of robotics.

No.	Laws of Robotics	List of Laws of Robotics
	David Langford Laws of Robotics (https://en.wikipedia.org/wiki/Ethics_robot%C3%B3w)	1. A robot will not harm authorized Government personnel but will terminate intruders with extreme prejudice. 2. A robot will obey the orders of authorized personnel except where such orders conflict with the Third Law. 3. A robot will guard its own existence with lethal antipersonnel weaponry, because a robot is bloody expensive.
	Asimov's Laws of Robots (Deng 2015, p. 25), (Asimov's Three Laws first appeared in the short story 'Playing in the Berk', later included in the anthology 'I, Robot', 6), on the basis of which the film of the same title, directed by Alex Proyas, starring Will Smith, was created in 2004.	1. A robot may not injure a human being or, through inaction, allow a human being to come to harm. 2. A robot must obey orders given to it by human beings except where such orders would conflict with the First Law. 3. A robot must protect its own existence as long as such protection does not conflict with the First or Second Law. 4. A robot cannot harm humanity, or by inaction cause harm to humanity.
	Mark Tilden's Laws of Robotics (https://en.wikipedia.org/wiki/Ethics_robot%C3%B3w)	1. A robot must protect its existence at all costs. 2. A robot must obtain and maintain access to its own power source. 3. A robot must continually search for better power sources.
4.	EURON (European Robotics Research Network) road map (http://www.roboethics.org/index_file/Roboethics%20Roadmap%20Rel.1.2.pdf)	1. Safety: Ensure human control of robot 2. Security: Prevent wrong or illegal use 1. Privacy: Protect data held by robot 2. Traceability: Record robot's activity 3. Identifiability: Give unique ID to each robot

Table 12 contd. ...

...Table 12 contd.

No.	Laws of Robotics	List of Laws of Robotics
5.	Friendly AI theory by Eliezer Yudkowsky (https://en.wikipedia.org/wiki/Ethics_robot%C3%B3w)	Requirements for Friendly Artificial Intelligence To be effective and to protect humanity from the unintended consequences of its actions, Friendly AI must satisfy the following conditions: 1. Friendliness – The AI must be benevolent toward humans and all sentient beings, making choices that are in the interest of all. 2. Preservation of Friendliness – The AI must be motivated to transmit its system of values to all its descendants, and to instill those values in similar intelligences. 3. Intelligence – The AI must be sufficiently capable of understanding how to pursue fairness through altruistic behavior and must strive to ensure that the results of its actions do not increase suffering. 4. Self-Improvement – The AI must feel the need and desire for continuous self-development and recognize and support that same desire for growth in the living beings around it.
6.	Approach proposed by Ray Wang (https://www.constellationr.com/blog-news/mondays-musings-designing-five-pillars-level-1-artificialintelligence-ethics)	The solution created should be: 1. Transparent – algorithms, attributes and correlations should be available for inspection by all interested parties; 2. Explainable – people should be able to understand how AI systems come to their contextual decisions; 3. Reversible – organizations must be able to reverse the learning process if needed (i.e., unteach an AI system of specific knowledge/behavior); 4. Trainable – AI systems must have the ability to learn from people and other systems; 5. Man-driven – all decisions should begin and end with human-made decision points.

Source: Own study.

To summarize this important aspect of the scientific discussion, it is necessary to refer in this approach to the broadly understood social context regarding algorithms, robots and cyborgs.

Kurzweil describes the relationship between humans and machines as follows: "The Singularity will represent the culmination of the merger of our biological thinking and existence with our technology, resulting in a world that is still human but that transcends our biological roots. There will be no distinction, post-Singularity, between human and machine or between physical and virtual reality. If you wonder what will remain unequivocally human in such a world, it's simply this quality: ours is the species that inherently seeks to extend its physical and mental reach beyond current limitations" (Kurzweil 2017, pp. 66–77, p. 25). It is therefore crucial to properly understand ethical issues and to transfer this understanding to well-defined and analyzed decision-making issues and mechanisms of assumed responsibility, along with attention to privacy and data security factors in this important component of artificial intelligence. Summarizing this aspect of the scientific discussion in legal terms, it should be noted that, for the time being, there are already a number of legal regulations regarding the design of artificial intelligence solutions and their legal exploitation. For example, the following have already been developed:

- OECD AI Principles – 2019 (https://www.oecd.org/en/topics/ai-principles.html),
- EU AI Ethics Guidelines for Trustworthy AI – 2019 (file:///C:/Users/Adam/Desktop/ai_hleg_ethics_guidelines_for_trustworthy_ai-en_87F84A41-A6E8-F38C-BFF661481B40077B_60419.pdf),
- UNESCO Recommendations on the Ethics of Artificial Intelligence 2022 (https://unesdoc.unesco.org/ark:/48223/pf0000381137, file:///C:/Users/Adam/Downloads/381137eng.pdf),
- EU Coordinated Plan on AI – 2021 (file:///C:/Users/Adam/Downloads/1_en_annexe_autre_acte_part1_v8_vf_C4B261EB-ABA4-5C30-1555482869410384_75787.pdf),
- Bletchley Declaration 2023 (https://www.gov.uk/government/publications/ai-safety-summit-2023-the-bletchley-declaration/the-bletchley-declaration-by-countries-attending-the-ai-safety-summit-1-2-november-2023),
- G7 Hiroshima Process on Generative AI – 2023 (file:///C:/Users/Adam/Desktop/bf3c0c60-en.pdf),
- NIST AI Risk Management Framework (NIST AI 600-1) – 2023 (https://nvlpubs.nist.gov/nistpubs/ai/NIST.AI.600-1.pdf),
- ISO/IEC 42001:2023 (standard for AI Management System (AIMS) – 2023 (https://www.iso.org/standard/81230.html),

- EU Artificial Intelligence Regulation (AI ACT) 2024 (https://eur-lex.europa.eu/legal-content/EN/TXT/PDF/?uri=OJ:L_202401689),
- OECD AI Policy Observatory (https://oecd.ai/en/).

These regulations draw attention to many elements related to the design and implementation of artificial intelligence solutions. At the same time, social and economic reality gives rise to more new scenarios, followed by further implemented regulations and formal and legal regulations.

It is therefore worth paying attention to the question of how to exist as human beings – individually, socially, collectively – in a world governed by algorithms (Benasayag 2021). In addition, one needs to be aware that various types of decisions based on AI algorithms can be fraught with numerous distortions resulting from, among others, inadequate data representation or built-in prejudices in algorithms.

4.14 Conclusions

Mechanisms and processes for the development and application of artificial intelligence for the purpose of creating positive climate change, including limiting climate hazards, are evolving, which is proved by the emergence of numerous concepts, the business models of companies and changes in the behavior of social groups. This is evident at the level of building and developing strategic information management systems in actions and crisis management systems for the climate and the neutralization of disasters, anomalies, dysfunctions and climate change. This type of action generates strong value by building climate security platforms at the level of various types of facilities, from companies to cities and countries.

References

Barfield, W., and Williams, A. (2017). *Cyborgs and Enhancement Technology, Philosophies*, 2, 4; doi:10.3390/philosophies2010004 www.mdpi.com/journal/philosophies.

Barfield, W., and Williams, A. (2017). *Law, Cyborgs, and Technologically Enhanced Brains, Philosophies*, 2, 6; doi:10.3390/philosophies2010006, s. 15.

Benasayag, M. (2021). *The tyranny of algorithms, transl.* S. Rendall, Europa, London.

Benyus, J. (2003). *Biomimicry*, HarperCollins, New York.

Bostrom, N. (2003). *Human Genetic Enhancements: A Transhumanist Perspective*. Journal of Value Inquiry, 37.

Chalmers, D. (2010). *The Singularity: A Philosophical Analysis*. Journal of Consciousness Studies, 17.

Chen, M. (2017). *Introduction to Cognitive Computing*. Huazhong University of Science and Technology, Wuhan, China.

Deng, B. (2015). *Machine ethics: The robot's dilemma*. Nature 2015, no 523, p. 25.

Di Francescomarino, C., Dumas, M., Federici, M., Ghidini, C., Maggi, F.M., Rizzi, W. et al. (2018). Genetic algorithms for hyperparameter optimization in predictive business process monitoring. Information Systems 74: 67–83, https://doi.org/10.1016/j.is.2018.01.003.

Dong, Y., Hou, J., Zhang, N., and Zhang, M. (2020). *Research on How Human Intelligence, Consciousness, and Cognitive Computing Affect the Development of Artificial Intelligence*, Hindawi Complexity, Volume, Article ID 1680845, 10 pages, https://doi.org/10.1155/2020/1680845.

Dorigo, M., and Stützle, T. (2004). *Ant Colony Optimization*. Bradford Company, Scituate, USA.

Eberhart, R.C., Shi, Y., and Kennedy, J. (2001). *Swarm Intelligence*. Morgan Kaufmann.

Erboz, G. (2017). *How To Define Industry 4.0: Main Pillars Of Industry 4.0*. Managerial Trends In The Development Of Enterprises In Globalization Era, Slovakia.

Espinoza, M. I., and Aronczyk, M. (2021). *Big data for climate action or climate action for big data?* Big Data & Society, January-June: 1–15, DOI: 10.1177/2053951720982032.

file:///C:/Users/Adam/Desktop/ai_hleg_ethics_guidelines_for_trustworthy_ai-en_87F84A41-A6E8-F38C-BFF661481B40077B_60419.pdf.

file:///C:/Users/Adam/Desktop/bf3c0c60-en.pdf.

file:///C:/Users/Adam/Downloads/1_en_annexe_autre_acte_part1_v8_vf_C4B261EB-ABA4-5C30-1555482869410384_75787.pdf.

Fox, P., and Hendler, J. (2014). *Science of data science: Big Data*, v. 2, no. 2, p. 68–70, https://doi.org/10.1089/big.2014.0011.

Glover, F. (1997). *Tabu Search and Adaptive Memory Programming – Advances, Applications and Challenges*. Operations Research/Computer Science Interfaces Series, 7: 1–75.

Goldberg, D.E. (2003). *Algorytmy genetyczne i ich zastosowania*. WNT, Warszawa.

Gollwitzer, P., Moskowitz, G. (1996). *Goal effect on action and cognition*. W: Higgins, E., and Kruglanski, A. (ed.). *Social psychology: Handbook of basic principles*, s. 361–399. New York: Guilford Press.

Habershon, E., and Woods, R. (2006). *No sex please, robot, just clean the floor*. The Sunday Times.

Hassani, H., Huang, X., and Silva, E. (2019). *Big Data and Climate Change, Big Data Cogn. Comput*, 3: 12; doi:10.3390/bdcc3010012.

Hayward, K. J., and Maas, M. M. (2021). *Artificial intelligence and crime: A primer for criminologists*. Crime, Media, Culture, 17(2): 209–233. https://doi.org/10.1177/1741659020917434.

http://www.roboethics.org/index_file/Roboethics%20Roadmap%20Rel.1.2.pdf.

https://eur-lex.europa.eu/legal-content/EN/TXT/PDF/?uri=OJ:L_202401689.

https://nvlpubs.nist.gov/nistpubs/ai/NIST.AI.600-1.pdf.

https://oecd.ai/en/.

https://pl.wikipedia.org/wiki/Cyborg.

https://pl.wikipedia.org/wiki/Etyka_robot%C3%B3w.

https://unesdoc.unesco.org/ark:/48223/pf0000381137,file:///C:/Users/Adam/Downloads/381137eng.pdf.

https://www. nature.com/news/machine-ethics-the-robot-s-dilemma-1.17881.

https://www.constellationr.com/blog-news/mondays-musings-designing-five-pillars-level-1-rtificialintelligence-ethics.

https://www.gov.uk/government/publications/ai-safety-summit-2023-the-bletchley-declaration/the-bletchley-declaration-by-countries-attending-the-ai-safety-summit-1-2-november-2023.

https://www.iso.org/standard/81230.html.

https://www.oecd.org/en/topics/ai-principles.html.

Jemielniak, D. (2025). Czwarte Prawo Robotyki [The Fourth Law of Robotics]. Nowa Fantastyka, 03(510): 16.

Księżak, P. (2021). *Prawo cyborgów (1)* [Cyborg Law (1)], Wprowadzenie w problematykę [Introduction]. Allerhand Law Review, Vol. 4 – No. 2 (8), item 29, e-ISSN 2657-800X.

Kurzweil, R. (2017). *Nadchodzi osobliwość* [The Singularity Is Near: When Humans Transcend Biology], Warszawa, pp. 66–77. p. 25.

Lem, S. (2020). *Summa technologiae*, Wydawnictwo Literackie.

Lepratte, L., and Yoguel, G. (2024). *Co-production, artificial intelligence and replication: the path of routine dynamics*. Rev. Evol. Polit. Econ. 5: 535–558, https://doi.org/10.1007/s43253-024-00134-0.

Tegmark, M. (2017). *Life 3.0, being human in the age of artificial intelligence.*

Mayer-Schönberger, V., and Cukier, K. (2014). *Big data: rewolucja, która zmieni nasze myślenie, pracę i życie* [Big Data: a revolution that will change our thinking, our work and our lives]. Warszawa: MT Biznes.

Michalewicz, Z. (2003). *Algorytmy genetyczne + struktury danych = programy ewolucyjne* [Genetic algorithms + data structures = evolutionary programs]. WNT, Warszawa.

Mitchell, M. (1995). *Genetic algorithms: an overview*. Complexity 1(1): 31–39 .

National Science Foundation (NSF). (2006). Cyber-Physical Systems (CPS), https://www.nsf.gov/pubs/2010/nsf10515/nsf10515.htm.

Neisser, U., Boodoo, G., Bouchard, T. J., Jr., Boykin, A. W., Brody, N., Ceci, S. J. et al. (1996). *Intelligence: knowns and unknowns. Am Psychol*. 51: 77–101.

NIST Special Publication 1500-201 Framework for Cyber-Physical Systems: Volume 1, Overview Version 1.0 Cyber-Physical Systems Public Working Group Smart Grid and Cyber-Physical Systems Program Office Engineering Laboratory, https://doi.org/10.6028/NIST.SP.1500-201.

Palwe, S., and Sirsikar, S. (2021). *Industry 4.0 Technologies and Their Applications in Fighting COVID-19*, [w:] Agrawal, R., Mittal, M., and Goyal, L.M. (eds.). Sustainability Measures for COVID-19 Pandemic, Springer, Singapore, s. 237–251. https://doi.org/10.1007/978981-16-3227-3_13.

Papadopoulos, T., and Balta, M.E. (2022). *Climate Change and big data analytics: Challenges and opportunities*. International Journal of Information Management 63: 102448, https://doi.org/10.1016/j.ijinfomgt.2021.102448.

Para, J., Del Ser, J., Aguirre, A., and Nebro, A.J. (2018). *Decision Making in Industry 4.0 Scenarios Supported by Imbalanced Data Classification*. In: Del Ser, J., Osaba, E., Bilbao, M. N., Sanchez-Medina, Vecchio M., and Yang, X.-S. (eds.). Intelligent Distributed Computing XII, Studies in Computational Intelligence, vol. 798, Springer, Cham, s. 121–134. https://doi.org/10.1007/978-3-319-99626-4_11.

Pransky, J. (2017). *The Pransky interview: Gianmarco Veruggio, Director of Research, CNR-IEIIT, Genoa Branch; Robotics Pioneer and Inventor*. Industrial Robot, Vol. 44 No. 1, pp. 6–10. https://doi.org/10.1108/IR-10-2016-0271.

Reis, M.S., and Geert, G. (2017). *Industrial Process Monitoring in the Big Data/Industry 4.0 Era: from Detection, to Diagnosis, to Prognosis, Processes*, 5(3): 35. https://doi.org/10.3390/pr5030035.

Rejmaniak, R. (2021). *Autonomiczność systemów sztucznej inteligencji jako wyzwanie dla prawa karnego* [Autonomy of AI systems as a challenge to criminal law], Roczniki Nauk Prawnych, Tom XXXI, numer 3, p. 95–113, DOI: http://doi.org/10.18290/rnp21313.6.

Różanowski, K. (2007). *Sztuczna inteligencja: rozwój, szanse i zagrożenia* [Artificial Intelligence: development, opportunities and threats], Zeszyty Naukowe 109–135, Zeszyty Naukowe Warszawskiej Wyższej Szkoły Informatyki, Warszawska Wyższa Szkoła Informatyki.

Sosińska-Kalata, B. (2018). *Big data (dane masowe) w nauce o informacji* [Big data in the science of information], ZIN, 56(2): 7–35.

Srinivasan, V., and Chakrabarti A. (2009). *Sapphire – an approach to analysis and Synthesis*, International conference on engineering design, iced'09, Stanford University, Stanford, CA, USA, ICED'09 2-417.

Sternberg, RJ. (2003). *Wisdom, Intelligence, and Creativity Synthesized*. New York, NY: Cambridge University Press.

Tadeusiewicz, R. (1993). Neural Networks [in Polish: Sieci neuronowe], Akademicka Oficyna Wydawnicza, Warszawa 1993 (ISBN 83-85769-03-X), (wyd. I i II).

Talwar, S., Kaur, P., Fosso Wamba, S. F., and Dhir, A. (2021). *Big Data in operations and supply chain management: A systematic literature review and future research agenda.* International Journal of Production Research, 59(11): 3509–3534.

Veruggio, G. (2005). *The Birth of Roboethics,* ICRA 2005, IEEE International Conference on Robotics and Automation Workshop on Robo-Ethics Barcelona.

Vincent, J.F.V., and Mann D.L. (2002). *Systematic Technology Transfer from Biology to Engineering.* Philosophical Transactions: Mathematical, Physical and Engineering Sciences, Vol. 360, No. 1791, Biomimetics: Technology Transfer from Biology to Engineering, pp. 159–173.

von Braun, J., Archer, M. S., Reichberg, G. M., and Sánchez Sorond, M. (2021). *AI, Robotics, and Humanity: Opportunities, Risks, and Implications for Ethics and Policy,* w: Robotics, AI, and Humanity, ed. J. von Braun et al., Springer, Cham, pp. 1–13.

Wamba, S., Akter, S., Edwards, A., Chopin, G., and Gnanzou, D. (2015). *How 'big data' can make big impact: Findings from a systematic review and a longitudinal case study.* International Journal of Production Economics, 165: 234–246.

Wang, H., and Wang, Y. (2022). Big data analytics for adaption to climate change, The 5th International Workshop on Environment and Geoscience, IOP Conf. Series: Earth and Environmental Science, 1087: 012014 IOP Publishing, doi:10.1088/1755-1315/1087/1/012014.

Chapter 5

Artificial Intelligence and an Overview of Selected Solutions for Creating Positive Climate Changes

5.1 Introduction

The aim of this chapter is to present artificial intelligence in relation to the review of selected solutions for creating positive climate changes.

The scope of the chapter covers the issues of creative artificial intelligence in terms of the climate, artificial intelligence technologies in the construction of digital business models of companies positively affecting climate change and generating a positive climate impact through generative design and intelligent algorithms.

The subject of the chapter concerns the approach to artificial intelligence from the perspective of creating and achieving a positive climate impact.

5.2 Climate change as a multidimensional global problem and artificial intelligence

Climate change should be perceived and interpreted as a strategic factor multiplying various types of hazards that amplifies existing trends and tendencies, tensions and situations of instability around the world. Climate change is a highly problematic issue that is ambiguously interpreted and understood. Many false theories and interpretations abound and much false information is spread on this very important topic. It is necessary to combat this and to explain what is untrue by means of solutions of the nature of scientific cognition. The issue of climate change is unique

DOI: 10.1201/9781003434016-5

because it is the essence of the global external effect and is an inherent problem of managing many public goods with great uncertainty. Climate change can only be combated through international cooperation due to the lack of an appropriate management structure over nation states (Jiang, Li 2020, pp. 165–171). Unfortunately, there is a lot to be worked out in this area, as different countries have different approaches and interpretations related to the fight against climate change. Climate change is primarily considered a man-made phenomenon that changes the way people live (Karanth et al. 2023). Current scientific discourses analyzing the problem related to climate change focus, among others, on the following issues:

- national security discourse – armed conflicts and climate change weaken the sovereignty and institutionality of the state,
- international security discourse – climate change poses a threat to international stability, emphasizing internationalism and global cooperation,
- human security discourse – climate change threatens human material conditions. The United Nations (UN) is an outstanding advocate of this discourse (Francisco 2023).

Ensuring sustainable living in the environment is essential for human beings. Today, we must pay attention, among others, to our local and global activities, for example, the use of fossil fuels, because the carbon footprint we leave threatens the entire planet (Houghton 2011, 14). The destructive impact of anthropogenic greenhouse gases generated during the use of fossil fuels and their contribution to climate change and global warming has been proven in many scientific studies and is accepted as a leading human and environmental problem of the 21st century (Tadesse 2010). The dynamic development of technology creates tools to fight global warming. The Internet of Things (IoT) and artificial intelligence (AI) have become the core technologies of these dedicated tools. Artificial intelligence can learn, establish relationships and make decisions about specific problems. Combining the capabilities of artificial intelligence with the practical use of the Internet of Things (IoT) would accelerate the fight against carbon emissions. The global implementation of these tools is increasingly significant in solving this important problem (Alpan et al. 2022, pp. 627–634). It should be added that the current political and social approach to combating climate change still needs to prove its ability to significantly reduce global greenhouse gas emissions in the near future. Therefore, research interest in climate engineering as a path to dynamic climate change has increased, strongly highlighting the place and role of technological and digital solutions in this process (The Royal Society 2009). Public acceptance of climate engineering strategies is not only important for the very reality of the deployment of these technologies; it is also necessary, at an initial stage, to carry out a number of experiments. Therefore, it is crucial for a better understanding of technologies to explore how they influence the attitudes, acceptance and behavioral intentions of managers of specific companies (Klaus, Ernst, Oswald 2020). Economic entities can therefore play a crucial role in climate protection. They can then be managed by applying the following aspects:

1. By shaping the attitudes of innovators to the implementation of innovation technologies and influencing the shaping of climate policy. This is because smart national governments play a crucial role in shaping climate policy, thus stimulating technological innovation and promoting business entities implementing technological innovation.
2. By implementing low-carbon technologies. In this area, interactions between technologies resulting from climate policy are crucial for future, emerging innovations and new market processes (Demertzidis et al. 2015, pp. 95–105).

Climate change acts as a hazard multiplier, i.e., climate change can exacerbate risks that already exist, especially in connection with the activities carried out by various actors for peace and safety (Sivapuram and Shaw 2020).

Since climate change is a global phenomenon, matching the size of an organization implies that some individuals may blame larger entities such as multinational corporations rather than individuals (Swim and Geige 2021). Scenario analyses are important in the climate change analysis process. Scenario-based management in relation to climate change is conceived as a tool for strategic planning and as a discourse tool to develop a common understanding of the subject among different actors. The concept itself does not focus on the specialized, individual needs of users, but rather opens up to a wide audience of various climate change scenarios for numerous social actors and policymakers (Auer et al. 2021). Awareness of climate change is the main driver of all business activities that focus on sustainable development. The dynamics of their internal changes are different and depend on the designed business models, strategies and application of specific technologies from economic processes (Arslan et al. 2021, pp. 720–730).

5.3 Creative artificial climate intelligence in climate change management processes

Creative artificial climate intelligence can play a special role in climate change management processes. Creative climate solutions should be a platform for creating an innovative approach to the development of artificial intelligence. In this interpretation, creative AI for the climate can

Regeneration Potential in Plants

Regeneration Potential in Plants delves into the intricate world of plant regeneration, exploring the biological and molecular mechanisms that enable plants to regenerate under different conditions. This book offers an authoritative perspective on one of the most fascinating and vital processes in plant biology, setting a benchmark for future research and applications. Unlike other volumes that focus solely on theoretical aspects, this comprehensive work bridges the gap between foundational knowledge and practical implications, showcasing its importance in fields ranging from genetic engineering to crop improvement.

Key features include:

- In-depth coverage of somatic embryogenesis and organogenesis as pivotal regeneration pathways.
- Detailed descriptions of signaling pathways, gene expression, and microRNAs involved in plant tissue regeneration.
- Insights into the latest biotechnological tools and techniques for enhancing plant regenerative capacities.
- Discussions on potential challenges and future directions in harnessing plant regeneration for sustainable development.

This book is an invaluable resource for plant biologists, researchers, and graduate students aiming to deepen their understanding of plant regeneration. Additionally, agricultural scientists and biotechnology professionals will find practical insights into the applications of plant regeneration for improving crop resilience and productivity. By blending high-level academic content with real-world examples, *Regeneration Potential in Plants* equips readers with the knowledge and tools to contribute to this dynamic and essential area of plant science.

be interpreted as a purposeful, dedicated technological tool for solving specifically defined climate problems. It may be based on a specially and continuously developed set of climate algorithms supporting and creating positive climate change in all possible ecosystems. The logic of these developed climate algorithms embodies a positive climate impact. Figure 17 shows the creative development cycle of an approach to managing climate change based on the use of artificial intelligence algorithms. The first cycle initiates the creation of successive cycles of creative development in an iterative way. After a certain number of cycles, the expected effect is created.

In this development cycle of a creative approach to managing climate change, the intertwining circulation of ideas and climate solutions generating the potential of creative capital for solving climate problems and managing potential and actual climate hazards is important.

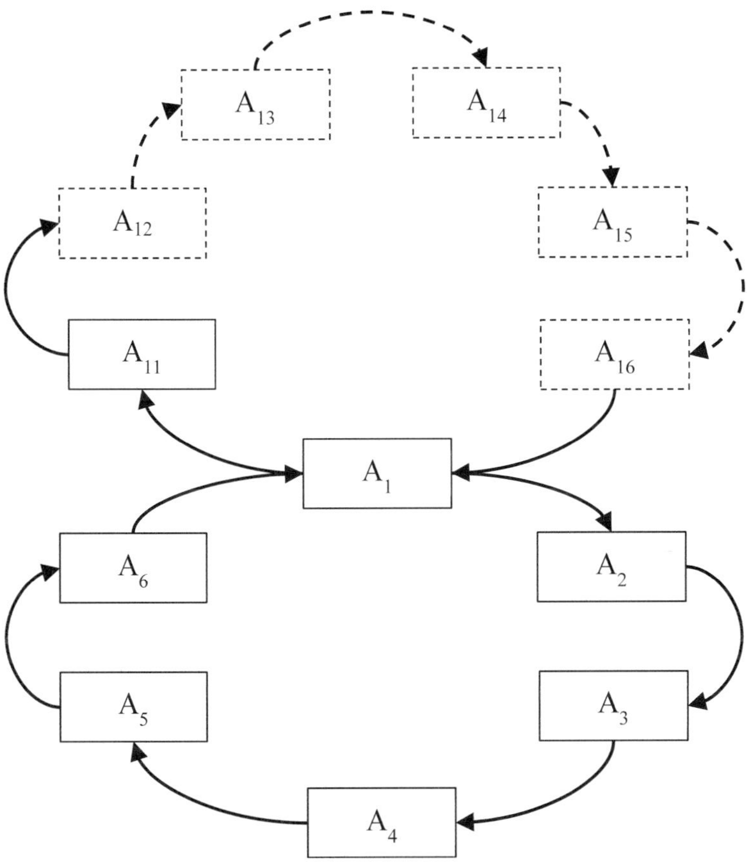

Figure 17. The development cycle of a creative approach to managing climate change.

5.4 Artificial intelligence in strategic climate information management

Artificial intelligence plays an important role in managing climate information. Access to climate information determines the activation of the decision-making process stream as a response to the climate information obtained. This can often also determine the preservation of human life or health if this information is related to climate hazards. This is related to the acquisition of information and its processing. Artificial intelligence can be used, among others, to process numerous data and a great amount of information in the so-called information environment. Therefore, the management of strategic climate information is an essential component of climate management. The lack of this information, or the lack of this information at the right time and place, for example, can potentially result in very difficult climatic phenomena of a catastrophic dimension. Figure 18 presents the concept of climate information management by means of artificial intelligence. Data on climate and climate change are collected in real time through different types of tools extracting these data from different sources. The data refer to all aspects that may be relevant to monitoring climate change.

In such a system of climate information management by means of artificial intelligence, the selection of appropriate and dedicated artificial intelligence mechanisms is most important in order to acquire climate information appropriately and to retrieve various types of data properly.

5.5 Artificial intelligence technologies in building digital business models of companies positively influencing climate change

Artificial Intelligence technologies open up the cognitive space for understanding many issues which have not yet been recognized or resolved. Their strength lies in as yet not fully discovered possibilities, also regarding the final results of its comprehensive application in business, the economy and society.

Across the world, artificial intelligence has become a hot topic in numerous scientific disciplines and in a number of ongoing public debates. Artificial intelligence is designed to simulate human intelligence to support and even expand human capabilities (Neubauer 2021). Intelligent technologies and machines are spreading in science and transforming the nature of the discovery process, and are now affecting the entire organization of science (Bianchini et al. 2022). The term 'artificial intelligence" was first used by the computer scientist McCarthy (McCarthy et al. 2006) in 1954 (Cukier 2019). At a conference organized by him and

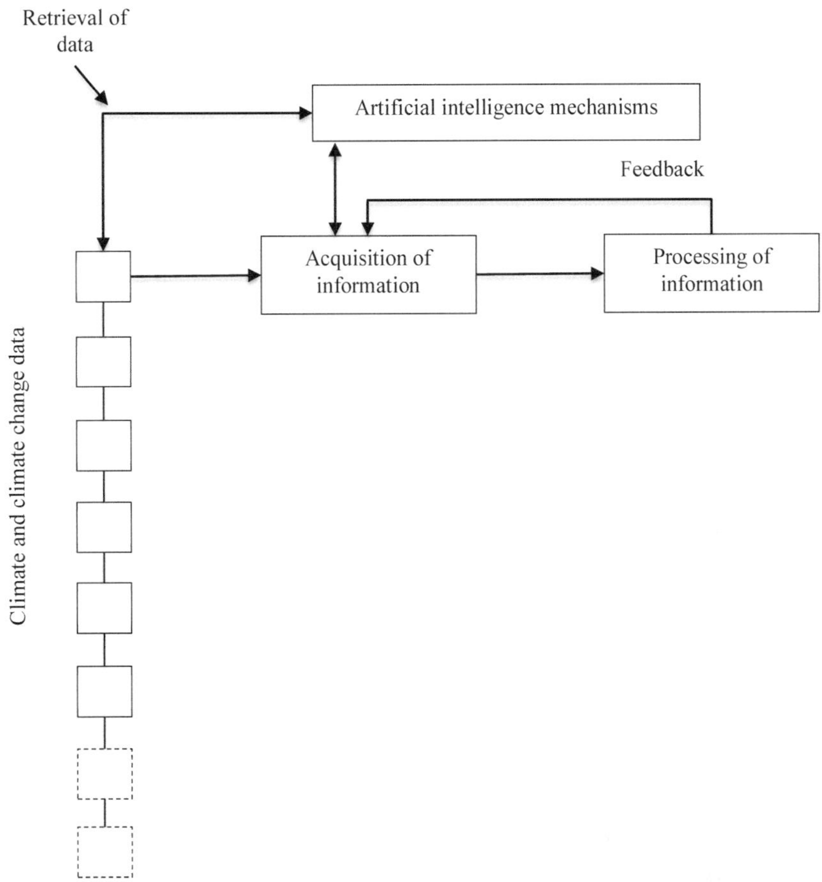

Figure 18. Climate information management by means of artificial intelligence. Source: Own study.

his colleagues, he stated that every aspect of learning and intelligence can be described in a way that can simulate a computer. Artificial intelligence is the ability to mimic human cognitive functions, such as learning and problem-solving, which are distinct features of the human mind (Schalkoff 1990). Artificial intelligence is a discipline that seeks to reproduce human ways of perceiving, reasoning, learning and solving problems and is an area of interest in practical and scientific applications (Reddy et al. 2019, pp. 22–28). Artificial intelligence as an intelligent subject can be divided into four areas: (i) thinking humanly, (ii) thinking rationally, (iii) acting humanly, and (iv) acting rationally. Artificial intelligence, as a system that works like a human being, is the field that has been most actively explored so far, and examples include natural language processing, automatic reasoning, knowledge presentation, speech recognition, machine learning, computer vision and robotics (Russell, Norvig 2017). Artificial intelligence has been identified as having three levels, namely artificial narrow intelligence (ANI), artificial general intelligence (AGI) and artificial super intelligence (ASI). Artificial narrow intelligence (ANI) is machines that are trained for a specific task and can only do it in one field. ANI is also known as weak AI, of which a typical example is Google search (Strelkova and Pasichnyk 2017, pp. 1–4). Artificial intelligence is a solution in which, for example, machines have the ability to think and make decisions as humans do, and a typical example of this is self-driving cars (Cully et al. 2015, pp. 503–507). Fuzzy systems and expert systems are popular tools for building artificial intelligence algorithms (Negnevitsky 2011). The use of fuzzy expert systems is justified where it is impossible to create a specific mathematical model describing inference. They are usually practiced as part of databases, control systems or operating on natural language. Fuzzy logic is very useful in engineering applications, i.e., where classical logic classifying only according to the criterion of true/false cannot effectively cope with numerous ambiguities and contradictions.

The concept of artificial intelligence is also a general term for a machine capable of perception, logic and learning. Machine learning is a subset of artificial intelligence and relies on the ability to learn without explicit programming. Deep learning is a subset of machine learning techniques and uses multi-layered neural networks of networks to build algorithms and apply them to solve a problem (Lee 2020). It is worth noting that the analysis of the Scopus database for titles, abstracts and keywords containing the terms “AI” and “artificial intelligence” also pointed to other keywords covering subfields of artificial intelligence, such as machine learning and various types of innovation, including product innovation, process innovation, business model innovation, incremental innovation, radical innovation, digital innovation, social innovation, sustainable innovation, open innovation, service innovation, breakthrough innovation, market innovation and organizational innovation (Mariani et al. 2023). There is, therefore, a strong relationship between artificial intelligence and the different types of innovations implemented by organizations, including in relation to climate change. This is due to the appreciation of technological management mechanisms taking into account climate technologies in various areas of business activity. With regard to the use of artificial intelligence in shaping sustainability mechanisms, certain skeptical approaches expressed in a holistic way are used. The issue of defining a methodology for calculating the relevance of the application of artificial intelligence to sustainable development principles, analyzing and evaluating the AI ethics and its relevance to the SDGs, as well as a framework for research work, projects and curricula, together with defining their relevance to the various SDGs arises (Nasir et al. 2023). Despite this, AI technologies are attracting increased interest, and are used in a number of research fields related to environmental protection, sustainable development and climate science (Galaz et al. 2021). Human activities are increasingly shaping the biosphere and climate system. Artificial intelligence and related technologies such as robotics and the Internet of Things (IoT), which are already able to increase the ability of societies to detect, adapt and respond to climate change and the environment (Campbell et al. 2019), are becoming a beacon of hope. Artificial intelligence also enables previously unfeasible activities to be carried out at a fraction of the cost of social management in energy-efficient cities (Ji and Huang 2022). AI-based approaches are not the "only tools in the locker" when it comes to dealing with the climate. Smart solutions can be offered through the use of artificial intelligence, from predicting and assisting people in developing sustainable decisions, to sustainable artificial intelligence for a socio-technical AI system with support for solutions for cyber-physical systems. On the other hand, unsustainable artificial intelligence puts the emphasis on AI applications and the integration of artificial intelligence into environmental and energy systems and other state-of-the-art technologies to facilitate the achievement of sustainability objectives (Saheb et al. 2022). In particular, it is difficult to overlook the contribution of blockchain (BT) and AI technologies to the era of the fourth industrial revolution (4IR). The former is rooted in the 4IR mechanism, while AI and BT have the potential to dynamically transform the current economic systems of enterprises (Polas et al. 2022). From this perspective, advanced artificial intelligence can facilitate the definition of so-called eco-climatic drivers. Different strategies for and approaches to mitigating man-made climate change, with a strong focus on mankind, society, the environment and values, should be taken into consideration (Taddeo et al. 2021, pp. 776–779). It is worth noting that we will never be able to fully return to the pre-digitalization state or an earlier model of human functioning. Even if attempts are made in this regard, knowledge already exists, it will expand and it will remain (Lamotte 2020). The proper implementation of artificial intelligence poses several challenges that need to be addressed in order to fully integrate it in a cost-effective way (Taddeo et al. 2021, pp. 776–779). First of all, the ethical problem arises, which is the basis of every other reflection in this field. The anthropocentric principle that artificial intelligence must always be put at the service of people, not the other way around, needs to be resolutely reaffirmed. The second challenge is technological: artificial intelligence is still not able to reproduce the complex functioning of the human mind, but only some of its limited capabilities. Another fundamental aspect concerns the skills that need to be developed in the age of artificial intelligence. The fourth challenge concerns the data which algorithms are based on. It is necessary for the data to be of good quality and as free as possible from bias due to errors in their generation. The last challenge is the legal issue.

In this regard, it is necessary to reconcile the principle of transparency and personal data protection with the right to privacy of the public (Felice 2022, pp. 1846–1856).

5.6 Digital business models based on artificial intelligence solutions that have a positive impact on climate change

The digital business model is a new and different category of business model. It has a different logic of generating income with a different way of allocating value to the market and a different shape and image of the value proposition offered. Managers of companies with a high impact on carbon emissions are currently under heavy pressure to develop business models and strategies to combat climate change. These strategies and business models take the form of a set of targets and components and plans to reduce greenhouse gas emissions and address changes in processes, markets and public order (de Abreu 2021). As regards building and implementing strategies, there is a need for an interdisciplinary and integrative approach to conducting research on business strategies so that climate change may gain increasing recognition among managers of different types of organizations. There is a consensus that such a cross-section of perspectives is currently lacking. Multi-level research methods on strategies and business models can satisfy this need due to their interdisciplinary nature (Paul et al. 2017, pp. 50–70).

Digital business models that have a positive impact on climate change can be built on the following assumptions:

- The co-creation approach is crucial for delivering innovation and works both offline (through increased competences within the network) and online (through the sharing of network-based products).
- Sectors with co-creation and a tailor-made approach are more effective. These include energy, water management and disaster risk reduction.
- Climate services mainly use the online "e-business model". Internet infrastructure is installed and updated as part of monthly, seasonal or annual subscription schemes (Larosa and Mysiak, 2020).

A digital business model with a positive impact on climate change can be called a progressive model. A progressive business model is not a consciously designed business model. Instead, it is the result of evolutionary interference by dictatorships and regulatory and legislative formulas that severely restrict business independence today. The progressive business model uses legislation/regulations to assign legal mandates and control important business decisions to implement a political set of climate criteria in the form of changing climate targets. It does not seek to control day-today operations, nor does it require the direct possession of facilities and other industrial resources. This model employs tactics that often evoke fear and moral indignation in all those who want to eliminate joint public debate on climate change (Rochlin 2021). In the context of shaping positive climate change, logic embedded in circular business models that also use digital transformation solutions can also be used. Circular business models are a way for corporations to stay competitive and solve sustainability problems. A typical example of such business models is the movement of enterprises towards the delivery of products as services, such as payment models for use, subscription or rental, with an emphasis on extending the lifespan and reliability of products, components and materials and/or reducing the overall consumption of the delivered products (Tukker 2004, pp. 246–260). Three business models can be used: the incumbent model (IBM), the pay-per-wash circular business model (CBM-PPW) and the circular business model product lease with lifetime extension (CBM-PLE) (Sigüenza et al. 2021, pp. 1084–1098). Climate-friendly digital business models have a great number of sustainable development goals, where their resilience and incentives to mitigate climate change are embedded. These include inclusive and adaptive business models.

An inclusive business model describes a way of creating, delivering and capturing value that provides access to resources (e.g., finance, technology, market channels) and space to make decisions about their use. The adaptive business model incorporates structure, processes and partnership principles taking into account a baseline approach based on a value pyramid, where the company effectively delivers and captures changes over time in response to internal or external shocks (Rosenstock 2020, pp. 76–81). The right choice of the type and structure of the digital business model always depends on managers. At the same time, it is important to know various typological and configuration solutions in order to choose a business model that will result in an acceptable, positive effect in relation to climate change.

5.7 AI business models and their application in climate management

In the new space of digital development, a new, very exciting and inspiring type of business model has emerged, namely the AI Business Model. The most interesting thing from this perspective is that such a business model can be called a technological representation in space and time of an appropriate configuration composed of components aggregated in artificial intelligence solutions for the development and growth of technologically intelligent enterprises. In the logic of a value-based business model, innovation redefines the boundary of business value or disrupts the current boundary of value. The innovation potential of a business model should be assessed on the basis of the logic of value creation and transformation. Based on this premise, a scheme with a set of hypothetical variables that help predict value changes in business model innovation is re-created (Chester et al. 2019). This is currently the case for business models based on artificial intelligence. Integrating artificial intelligence into business models means a significant evolution in how companies operate and develop strategies. AI-based business models are characterized by the integration of AI technologies to improve, innovate or create new aspects of a company's business, products or services. AI-based business models are distinguished by the ability to use data and machine learning (ML) technologies to create more efficient, responsive and intelligent business processes (Farayola et al. 2023). In the area of customized production, AI-based models are changing traditional production paradigms. A customized AI-based manufacturing factory is characterized by self-assessment, operation optimization, dynamic reconfiguration and intelligent decision-making capabilities. These features highlight the transformative impact of artificial intelligence on production, moving towards more flexible and efficient production methods (Wan et al. 2020).

The archetypal business model patterns of AI startups (Weber et al. 2022) are as follows**:**

1. **AI-charged Product/Service Provider** – Provide products and services that have embedded AI models embedded.
2. **AI Development Facilitator** – Facilitate AI development of customers with customizable solutions or technical interfaces.
3. **Data Analytics Provider** – Provide solutions that integrate and analyze various data sources for decision support.
4. **Deep Tech Researcher** – Research and develop AI based on creating technology for supporting innovative problems.

The AI business platform model is practically on par with the cloud-based SaaS model. This includes AI solutions that can work together on top of other digital systems, such as CRM (Customer Relationship Management) and ERP (Enterprise Resource Planning) business systems. AI boundary data flows to digital data through coordination, driving business improvements in phases. In this business model, the company will securely store a recurring subscription (Mishra and Tripathi 2021). Theoretical concepts in the construction of AI Business Models are related to research into AI, BMI, digital transformation and business ecosystems. Therefore, core conclusions are defined as the need to (1) understand AI and organizational capabilities needed for digital transformation; (2) understand the current BM, the potential for BMI and the role of the business ecosystem; (3) develop and improve the capabilities needed for AI deployment; and (4) achieve organizational acceptance and develop internal competencies (Reim et al. 2020).

5.8 Positive climate impact as a result of the use of digital business models and strategies based on artificial intelligence solutions

It is not an easy task to achieve a positive climate impact. First, you need to answer the question of where this impact is to be achieved, at the level of, e.g., Europe, a single country, a region, a city, a commune or an enterprise, in the area of international corporation with a global scope or a locally operating enterprise. Subsequently, a focus on how this impact should be expressed in terms of the market, and then how to measure it by means of the measurement and monitoring aspects, is necessary. The proper design and implementation of digital business models based on AI solutions shaping the positive impact on climate change can result in new, innovative climate services being delivered through these business models. As the impact of climate change becomes increasingly apparent, various types of innovation are expected to play a central role in enabling sustainable development at the national level and rapid decarbonization. Recent events around the world present the technological approach of both private companies and public sectors to the use of low or carbon-free technologies limiting sources of greenhouse gases (GHG), to innovation in carbon capture and storage, so as to effectively address the effects of global warming (Matos et al. 2022). To improve the effectiveness of positive climate change, climate service providers should understand the different links in the value chain (or network), identifying core actors, and engage or cooperate to jointly develop climate services valuable to the market and society. In fact, all actors along the value chain would benefit from a better understanding of all links in the value chain stream (Hewitt and Stone 2021). It is important that company boards take into account dynamic climate changes so that they are part of their responsibilities as managers. Integrating climate change issues into strategic and supervisory responsibilities can also be helpful. These organizations are then more likely to take into account the risks and opportunities of climate change in their decision-making processes (Albitar et al. 2023). The positive impact on climate change is closely linked to the use or change of technology. There is a strong relationship between climate policy and technological change induced in environmentally friendly technologies. Research results and expert analyses show that a stricter emissions cap means (i) lower use of fossil fuels and incentives to improve the efficiency of fossil fuels; (ii) higher incentives to improve the efficiency of alternative energy; (iii) a positive impact on direct carbon reduction technologies; and (iv) lower overall incentives to innovate (Wang 2021). From this cognitive perspective, a strong synergistic effect of the entire economy for positive climate change occurs. The use of artificial intelligence in shaping digital business models that have a positive impact on climate change is a considerable challenge both at the levels of the entire economy and specific companies. This is, among others, related to the use of AI technologies in shaping the digital business models of companies, different interpretations of digital business models based on AI solutions positively affecting climate change and the achievable climate impact through digital business models based on AI solutions. The situation is similar with regard to digital climate strategies, which can be defined as a set of long-term goals of organizations focused on the principles of the digital and climate transformation, i.e., a set of digital climate goals. Digital climate targets will have a hybrid function, i.e., it will be a platform based on fulfilling the principles of cumulative digital and climate value aggregated into one coherent system of strategic objectives. However, it should be remembered that we currently exist in a market situation of permanent unpredictability and a total state of chaos. Hence, the key is the implementation of emergent strategies, generated through various observed and used types of opportunities appearing in various places of the socio-economic ecosystem. Moreover, strategic actions are supported by actions at the tactical–operational level called stractics.

5.9 Generative design in climate change management

Generative design consists of an action in which the constructor, the designer, precisely determines the input data for the design process being implemented, such as material type, weight, strength, production technologies and product price. Based on boundary conditions defined as such, the computer uses dedicated algorithms based on artificial intelligence to create thousands of different concepts and analyze and verify each of them. This is one of the core stages of using the generative design function, where we define specific boundary conditions ourselves. An important functionality which is useful when designing by means of artificial intelligence is behavioral modeling, which uses neural algorithms to analyze the behavior of a current model and predict its response to changes in parameters or specific conditions. In the case of generative design, we are talking about the so-called AI-Driven Design, a specific solution with a high degree of autonomy that can independently generate and optimize projects, acting as the main "constructor" (Jenis et al. 2023). Such solutions can be widely used in climate change management processes.

Examples of generative design solutions for positive climate change may include the following:

1. Creating digital twins of current climate projects.
2. Creating 3D models of construction projects for sustainable climate construction.
3. Optimizing energy use in buildings that meet climate criteria.
4. Project documentation management by means of artificial intelligence.

The impact of artificial intelligence on sustainable climate construction can be achieved through a multidimensional combination with Building Information Modeling (BIM) and Virtual Reality (VR) technology. Then, technological progress is significant and leads to a huge technological revolution in the construction and investment industry. The extensive integration of artificial intelligence with BIM and VR technology can dynamically contribute to the sustainable design of buildings, the analysis of their energy efficiency, the enhanced safety of sustainable infrastructure and optimized construction project management processes themselves. Thanks to such advanced AI algorithms, architects, engineers, technologists and constructors can create more efficient, sustainable and safe buildings, which can definitely contribute to improving the quality of life of society, protecting the natural environment and generating, in essence, a number of positive climate changes for the economy and society.

5.10 Smart algorithms and their impact on the design of free natural space

Intelligent algorithms can be used to design free natural space as places of new human existence in harmony with nature. The return to nature through the use of artificial intelligence solutions is a very important and fundamental area of these applications in the practice of generating positive climate change. The principles of urban bioclimatic design apply here. Then, the natural environment of the city can be an area of renewal, protection and a new purpose and scope of making urban space available to the population as a new ecological and climate space. Such a space is friendly to nature and the climate. In addition, a new place and the role of spatial planning in environmental and climate management are vital here. Related projects satisfy the criteria relevant to ecophysiographic studies, understood as documentation prepared for the purposes of the study of the conditions and directions of spatial development of the municipality, the local spatial development plan and the spatial development plan of the province, characterizing individual natural elements in the area covered by the study or plan and their interrelationships. In this context, universal design is important. In addition to universal design, the relevant literature comprises concepts such as inclusive design and Design for All. Such design is comprehensive and systemic, strongly strengthening knowledge and strategic climate information.

5.11 The overview of selected artificial intelligence solutions for creating positive climate changes

The use of artificial intelligence in the process of managing climate hazards is becoming, on the one hand, a goal and priority and, on the other hand, a crucial intellectual strategic challenge. Artificial intelligence models are a dynamic engine that drives innovation. These generative models can use computer vision, natural language processing, and machine learning to analyze complex data patterns. For example, ChatGPT, created by OpenAI (the work was financed by Microsoft), Bard and Gemini, created by Google, and Grok, developed by Elon Musk's start-up xAI, are artificial intelligence tools that can generate shorter or longer answers to any questions asked. In addition to pattern recognition, AI models for learning also use decision-making algorithms. Through the process of training, collecting and analyzing data, they are coming ever closer to mastering the activities and goals set for them. One scientific study analyzed the occurrence of energy-related keywords by describing AI and ML in articles. It was discovered that the current literature can be divided into four main groups, namely (1) AI applications in performance and use; (2) ML for forecasting; (3) algorithm and pattern recognition for learning systems; and (4) management and transport of energy sources.

The inclusion of important energy domains in our analysis has shown that AI and ML are currently used in a small number of energy-related domains and have promising potential for integration with other sections. This study has revealed many promising and neglected areas where AI can be deployed, including energy storage, uncertainty analysis, wastewater treatment, emission control, biofuel production, the energy supply chain, renewable energy sources, risk assessment and demand response. As regards ML, in addition to the unexplored areas listed in the sections above, heat storage, heat energy, carbon capture techniques, environmental management, economics and investments, renewable energy and uncertainty analysis can be identified (Entezari et al. 2023). Seasonal Auto-Regressive Integrated Moving Average (SARIMA), Long Short-Term Memory (LSTM) and Gated Recurrent Unit (GRU) models are used to predict temperature until 2050 based on historical data (Reine Nishimwe and Reiter 2022).

AI technologies can effectively address problems related to the integration of renewable energy into the energy system, such as solar and wind energy forecasting, frequency analysis and control of the energy system, and transitional stability assessment (Zhang et al. 2022). For example, the explainable artificial intelligence (XAI) model has been developed to analyze the impact of climate change on the consumption of cooling energy (Ec) in buildings, to predict the long-term Ec under new common climate change scenarios of shared socioeconomic pathways (SSP) and to explain the underlying causes of the projections. Such analyses and future projections are necessary to enable decision-makers and stakeholders to achieve change-resistant climate and sustainable development objectives by harnessing the power of meaningful and reliable projections and insights (Chakraborty et al. 2021). Artificial intelligence is also used in communication networks. The 6G communication network will be the next generation of the sixth sense communication network, which will increase the value of the intelligent Internet of Things. With the advent of various fields of artificial intelligence, 6G will create huge opportunities, that is, the growth of human intelligence, the Internet of Everything, the quality of experience, the quality of life, etc. Artificial intelligence and 6G communication technology will completely change from connected things to connected intelligence (Ahammed et al. 2022). The agent-based AI tool is a very useful instrument for assessing future scenarios where different alternative solutions and approximate outcomes can be assessed that are closer to the expected outcomes. As regards the analyzed projects, it can also promote the development of sustainable agriculture (Sánchez et al. 2022). Artificial intelligence can also be used in the construction sector. In the pursuit of a low-carbon society, it is essential to increase the feasibility and applicability of AI techniques in the construction sector, as deep learning building load forecasting can improve the energy efficiency of buildings. However, a large amount of sensor data is required to be used in the development of a building load forecasting model based on deep learning. Reducing the costs of detecting and pre-processing weather and building data would make stakeholders or building owners more interested in using deep learning-based predictive building energy control systems. To promote the applicability of the deep learning method in buildings, it is therefore important to identify the most influential input variables, thus creating a cost-effective deep learning building load forecasting system (Chung and Liu 2022). The power management model based on artificial intelligence (AI-EMM) has a strong application in green construction. Adaptable to human choices, it can act intelligently to increase user comfort, safety and energy efficiency. One of the core elements of the AI-EMM model is a universal infrared communication system and subsystems for intelligent user identification and monitoring of the internal and external environment. Long short-term memory models (LSTM) are used to increase energy consumption. Data on energy consumption in a green building are analyzed by means of the suggested approach. In order to achieve a better indoor climate, research investigating the relationship between heating, ventilation and air conditioning (HVAC) systems should focus on optimizing the design on the air side. According to the findings, the economic benefits and ecologically sustainable building coexist harmoniously. A green or sustainable building is one in which the structure and features protect or improve the local environment (Xiang et al. 2022). Steam cooling systems with a dehumidifier pave the way for sustainable energy and environmental development in buildings in particular; however, direct steam coolers in such configurations result in high water consumption. The use of modern computational intelligence tools, including artificial intelligence and metaheuristic optimization algorithms, can improve the operational understanding of cooling systems with a dehumidifier, while dealing with minimizing the total water footprint while maximizing cooling efficiency (Tariq et al. 2023). Artificial intelligence has become an important area for addressing most environmental sustainability issues, such as biodiversity, energy, transport and water management. Biodiversity research has led to the development of machine learning or natural language processing solutions to predict ecosystem services. Artificial intelligence applications and machine learning models are increasingly being used to predict and optimize the protection of water resources. Area neural networks, expert systems, pattern recognition and fuzzy logic models are the main areas of concentration in energy. Applications of computer vision and decision support were found in transportation. Timely monitoring of interventions is required to improve environmental sustainability (Nti et al. 2022). Human-induced carbon emissions and the consequent increase in global warming are major threats to our livelihoods. Although there are many causes of carbon dioxide emissions, the energy used by homes also has a significant impact on this. The development of technology creates tools to fight global warming. The Internet of Things (IoT) and artificial intelligence (AI) have become essential tools. Artificial intelligence can learn, relate and make decisions about specific problems, and combining AI capabilities with the practical and efficient use of IoT would accelerate the fight against carbon emissions. However, the global implementation of these tools is increasingly important for the solution (Alpan et al. 2022). The transition to carbon neutrality in construction sectors requires a combination of renewable systems and artificial intelligence to ensure robustness, reliability, automation and flexibility. AI techniques can be used in renewable energy for solar potential assessment, multi-level stochastic uncertainty analysis, intelligent control, fault detection and diagnosis, and single- and multi-purpose optimizations. The results showed that compared to physical models, data-driven models demonstrate superiority in terms of modelling simplification and modification, high prediction accuracy and computational performance.

To improve the operational robustness of the system under the uncertainty of the tiered scenario, a general approach with artificial intelligence is proposed, including non-linear development of the data-driven model, energy efficiency forecasting based on uncertainty, stochastic uncertainty and statistical analysis. In addition, artificial intelligence can also be used in renewable systems to ensure safety, reliability and stability (Zhou 2022). Hybrid Energy Systems (HES) combine multiple energy sources to maximize energy efficiency. Due to its unpredictability and dependence on weather, the integration of renewable energy sources (RES) is a viable option for distributed generation (DG). AI techniques are widely used in HES, and the study cited was on how artificial intelligence can solve problems of classification, forecasting, networking, optimization, and control. This study provides an overview of the recent history of HES' critical challenges in energy management, dimensioning, demand management and storage management; in addition, several conceptual/theoretical problems, predecessors and consequences that may be of interest or require further research have been addressed. Companies must ensure that their systems operate effectively and repay their investments. Regardless of the system, failures and defects should be diagnosed and repaired as soon as possible. This can be achieved by increasing system performance and preventing damage at an early stage (Maghami and Mutambara 2023). The development and implementation of AI technologies for healthcare is a growing field. Therefore, it is possible to ensure that the climate impact of these technologies is taken into account and considered throughout the life cycle of development and deployment. Based on current research, NHS England needs to develop and implement a formal methodology to assess the climate impact of AI technologies used in the healthcare system. While there is great potential for the use of artificial intelligence to improve healthcare delivery, surgeries and population health, more research is needed to find out where and how these tools could be most effective in reducing NHS-related emissions (Bloomfield et al. 2021). New machine learning and deep learning techniques are constantly being proposed to improve performance to reduce emissions. The emergence of AI models and advanced algorithms ensures the optimal use of computational resources in healthcare, contributing to faster processing of results and monitoring, thus minimizing the uptime of different machines and healthcare systems, leading to emission reductions (Das and Chandra 2023). Climate variables play an important role in increased tree ring width (TRW), which is one of the main paleoclimatic signals. This study uses new methods to understand the response of tree growth to climate variables under the influence of anthropogenic climate change. This study used AI regression and modelling techniques to develop a model describing the relationships between climate variables and the tree ring standardized growth index

(TRSGI) in northeastern South Korea. Data from 1901 to 1998 were analyzed and then data from 1999 to 2019 were used to reconstruct the TRSGI (Salehnia and Ahn 2022). Growing crops with high yields and better adaptation to new and changing climates is necessary to ensure continued food security, biomass production and ecosystem services. Advances in genomics and phenomics provide insight into the complex biological mechanisms that underlie plant function in response to environmental disturbances. However, linking genotype to phenotype remains a huge challenge and hinders the optimal application of high-throughput genomics and phenomics in advanced breeding. The key to success is the need to assimilate large amounts of data into biologically meaningful interpretations. The current state of field genomics and phenomics allows for the exploration of emerging approaches to and challenges of integrating large datasets in the field of ultramicro-omics with the help of next-generation AI (Next-Gen) and proposing practical paths for improvement (Harfouche et al. 2019). Artificial intelligence and machine learning are increasingly used for prediction in agricultural science. However, many models are usually black boxes, which means that we cannot explain what the models have learned from the data and why the forecasts are the same (Ryo 2022, pp. 257–265). Artificial intelligence can also be used for the emerging, digitally driven management of the Earth's system for the protection of marine biodiversity. Artificial intelligence is widely used in agriculture. To grow healthier crops, fight pests, monitor soil and growth conditions, analyze data for farmers and improve other management activities of the food supply chain, the agricultural sector is turning to AI technologies. This makes it difficult for farmers to choose the perfect time to plant seeds. Artificial intelligence helps farmers choose the optimal seeds under certain weather conditions. It also offers data on weather forecasts. AI-based solutions will help farmers produce more with fewer resources, increase crop quality and speed up the time at which products reach the market. Artificial intelligence helps to understand soil quality (Javaid et al. 2022). The growing population and the effects of climate change have placed enormous responsibility on the agricultural sector to increase crop production and productivity. In most countries where arable land expansion is simply impossible, agricultural automation has become the only option and is the need of the moment. The Internet of Things and artificial intelligence have already begun to capitalize in all industries, including agriculture. Advances in these digital technologies have made revolutionary changes in agriculture, providing intelligent systems that can monitor, control and visualize various agricultural operations in real time and with the comparable intelligence of human experts. Potential applications of IoT and AI can be used in the development of intelligent agricultural machinery, irrigation systems, weed and pest control, fertilizer use, greenhouse cultivation, storage structures, crop protection drones, crop condition monitoring, etc. (Subeesh and Mehta 2021, pp. 278–291). Although they generally receive little attention in vegetation research, ground lichens (terricolous) are the main contributors to the overall carbon and nitrogen cycle, albedo, biodiversity and biomass in many high latitude ecosystems. Changes in the biomass forming the mat of bright lichens can affect vegetation, fauna, climate and human activities, including reindeer breeding. Lichens have a complex spectral signature, and terricolous lichens have a limited height of growth, often growing in mixtures with higher vegetation. So far, this has prevented the development of remote sensing techniques to accurately assess lichen biomass, which would be a powerful tool in ecosystem and ecological research and pasture management (Erlandsson et al. 2022). Artificial intelligence and automation can help fungi-based bioprocesses increase production efficiency. Machine learning can uncover nonlinear interactions when growing mushrooms.

Soft sensors make it possible to understand the complex dynamics of a bioprocess. Robots minimize direct labor costs while improving the quality of mushroom products. Further improvements are needed to automate industrial fungi-based bioprocesses (Wainaina and Taherzadeh 2023). Artificial intelligence can analyze air data. One study assessed the performance of an artificial neural network (ANN) architecture with sigmoid neurons in a hidden layer, trained using an evolutionary algorithm (evolutionary ANN, EANN). This architecture of the XAI model (XAI-EANN) has been applied to the long-term forecast of air temperature in various subregions in the south of the Iberian Peninsula. In this case, the average air temperature in August was predicted on the basis of ERA5 reanalysis variables, obtaining good forecasting skills and explainable models for the climatological variables under consideration. First, a cluster analysis was carried out for the average air temperature in the zone, in such a way that a number of subregions with different air temperature maintenance were defined (Gómez-Orellana et al. 2023). Mobile Marine Protected Areas (MMPAs) are based on artificial intelligence. MMPAs are geographically dynamic, with moving boundaries that change location as endangered species migrate across the ocean. This real-time, mobile and potentially spatially ubiquitous form of ocean management relies on digital equipment that collects data from a variety of sources (e.g., nanosatellites, drones, environmental sensor networks, digital bioacoustics, marine tags, deep-sea unmanned aerial vehicles), combined with analytics such as machine learning algorithms, computer vision and green computing techniques. MMPAs are justified on the basis of their ability to enable responsive, real-time adaptation to environmental variability, species mobility and disturbance dynamics; therefore, scientists and regulators are increasingly recommending the deployment of these AI-based computing systems in the world's oceans

(Bakker 2022). The mobility sector is undergoing a global transformation towards cleaner and more sustainable technologies. Many mobility projects are developing AI technologies to improve the operational efficiency of mobility, such as optimizing the charging system for electric vehicles (EVs), autonomous driving and traffic control. This study presents the global distribution of such projects, showing a map generated by the Geographic Information System (GIS) and analyzing the level of readiness of these technologies, using the Japan Technological Readiness Assessment (J-TRA) methodology. The results show that most of the projects are located in Europe. Among the AI applications analyzed, intelligent parking systems and lane-tracking assistive technologies have the highest level of readiness. Further AI training to ensure full compliance with real-world operational environments and updates to traffic rules are necessary to enable advances in the technological readiness of other types of AI mobility technologies (Pandyaswargo et al. 2023, pp. 720–727). EMU means electric multiple units and refers to a train of self-propelled electric wagons. Energy from renewable sources such as solar and wind energy can be stored in energy storage systems (BESS) and released when consumers need it most. Electric traction motors are built inside one or more EMU wagons, eliminating the need for a separate motor. Electric trains can go further and faster than those powered by steam engines. There are several advantages to using electric trains, such as the efficient conversion of fuel to kinetic energy. Rail travel is affected by fluctuations in electrical power. With artificial intelligence, energy waste and prices can be reduced, and the deployment of clean, renewable energy sources can be facilitated and accelerated in energy systems around the world. Artificial intelligence can benefit from the design, operation and management of energy systems (Liu et al. 2022). Based on classified AI algorithms, this study shows a dynamic, changing pattern of convergence between AI and electric vehicle technology and reveals how AI has impacted innovation in electric vehicle technology over time. This study reveals that artificial intelligence is accelerating the automation of electric vehicle driving, and that artificial intelligence algorithms that are widely used in electric vehicles have changed over time, and the technological areas of electric vehicles affected by AI have also changed (Lee 2020). So-called algorithmic biases appear in effective management with the use of artificial intelligence. Whether due to inappropriate training data, inappropriate contexts or user interpretation errors, algorithmic biases are common and must be carefully considered in the field of sustainability. In the fields of agriculture, environment and sustainability, such biases may result, for example, in risks to critical elements of food security and ecosystem resilience.

Crucial questions for the future are as follows:

- To what extent do insights on and risk management solutions to algorithmic biases from other domains apply in sectors such as digital agriculture, digital forestry, urban planning and marine resource extraction and management?
- How are the predictive potential and effectiveness of AI models affected by the fact that ecosystems, such as the terrestrial and marine landscapes, are changing rapidly due to, for example, climate change?
- What social, economic and environmental impacts can arise from these biases and how should they be prevented? (Galaz et al. 2021).

Precision farming, including the deployment of robotic farm workers, AI-powered equipment and relevant smart systems, is enthusiastically praised for improving crops, strengthening food security, generating economic growth and fighting poverty. Techno optimism has captured the imagination of the media, industry and governments. At the same time, computer science and machine learning researchers have begun cataloguing the potential harms of information technologies that are shaped by prejudice, discrimination, and Western power hierarchies. While precision farming and smart farming technologies can provide some opportunities for farmers, they can also emerge as a new, albeit familiar, system for appropriating and controlling their work and knowledge (Foster 2023). Modern breeding programs face the challenges of collecting, archiving and extracting Big Data from multiple sources. The raw data generated in plant breeding must be processed into a format that can be used for computer analysis (e.g., machine learning). Data processing includes cleaning (identifying and replacing incomplete, inaccurate, irrelevant or otherwise problematic data), integration (combining data from different sources and formats), transformation (performing standardization or hierarchy of concepts to properly format data), reduction (presenting a less complex representation of data, e.g., through dimensionality reduction (DR), data abundance reduction and compression), discretization (replacing continuous raw values with interval ranges) and sampling (selecting a subset of samples enabling the estimation of the entire population). After being processed, data can be explored by means of statistical analysis, ML, artificial neural networks (ANNs) or pattern discovery. Some of these methods are used in modelling and forecasting (Xu et al. 2022, pp. 1664–1695).

5.12 Artificial Intelligence in shaping digital business models that positively impact climate change

The application of appropriate dynamics when making changes in the climate area is fundamental due to the strength of hazards determined by these changes, because the dynamics of social and economic changes are related to the overlapping of the stream of numerous crises creating a dangerous environment in which various types of organizations and individuals operate and/or have had to operate in the last few years.

The environment of many crises, including economic, social, political, the climate crisis caused by significant human interference over many years in the structure of the natural environment, the health crisis caused by the COVID-19 pandemic and the military and migration crisis caused by the war in Ukraine can be neutralized by organizations by building a common platform, creating an environment of numerous business models. Due to the above-mentioned crises, the digital and climate transformation has significantly accelerated, affecting a wide area of business and human existence at local, national and global levels. In this space, companies, by eliminating all types of threats, are searching for opportunities that will ensure their ability to survive even in the most unfavorable conditions. They see such opportunities, among others, in the use of new technologies, which include artificial intelligence. Artificial intelligence is now becoming a source of shaping digital business models, which are a crucial tool for capturing value from a challenging and economically constrained market. This is done through a value proposition offered and delivered to customers by means of analog and digital solutions (in analog terms to direct service customers, including citizens) and digitally for Internet users and netizens. These business models are also implemented through the development and application of such an income generation logic that can provide companies, in the short and long term, with economic surpluses to protect business from adverse phenomena and maintain the ability to develop them. Artificial intelligence solutions are at the forefront of this socio-economic ecosystem. They are built, among others, in digital business models with the use of technology based on algorithms interacting and creating new learning processes of humans and humanoids, while also affecting the mechanisms of processing thinking processes from a machine perspective, taking into account the aspects of automation and robotics. It should also be noted that with the diametrically changing value systems of societies, an important factor determining the success of the organization in the current socio-economic conditions is existential security, including economic security and the quality of life of people based on the search for balance in many spheres of their functioning. Thus, the changing, mainly adverse climatic conditions take on particular importance. Not only do they have a strong impact, but in many parts of the world they completely change the way in which the basic functions of local societies are carried out. Artificial intelligence plays a role in shaping digital business models that positively affect climate change. Digital business models are constructed and implemented as a complex platform. This can serve not only to reduce the adverse impact on the environment, but above all it can generate, preferably abruptly, a positive impact on climate change. This can be achieved through the use of socio-technological instruments built into these digital business models. The scope of the content is not only local, but mainly global, i.e., one that will allow the organizational, managerial and technical–technological solutions developed in companies using digital business models that positively change the climate to be extrapolated to different regions of the world in different configurations and cognitive spaces. The authors ask the following research questions, which they try to answer through the use of multidimensional scientific discourse:

1. How can AI technologies be used to shape digital business models for businesses?
2. How should the digital business model based on AI solutions positively impacting climate change be interpreted?
3. What positive climate impact (of what nature) can be achieved through digital business models based on AI solutions?

The authors try to answer the above questions using a complex scientific discourse with a wide use of the available relevant global literature.

5.13 Conclusions

The adoption of appropriate and optimal principles for building climate security should be supported by all possible AI solutions. This is especially important in taming the dynamics of anomalies and climate dysfunctions in real time. The logic of smart climate management requires a number of streams of the intelligent use of AI to mitigate climate hazards. Therefore, strategic decision-makers must have climate awareness and sensitivity based on mindfulness and empathy and strategic reflection related to properly interpreted climate hazards in the processes of human and economic existence

References

Ahammed, T.B., Patgiri, R., and Nayak, S. (2022). *A vision on the artificial intelligence for 6G communication*. ICT Express, https://doi.org/10.1016/j.icte.2022.05.005.

Albitar, K., Al-Shaer, H., and Liu, Y.S. (2023). *Corporate commitment to climate change: The effect of eco-innovation and climate governance*. Research Policy. 52, 2, 104697, DOI 10.1016/j.respol.2022.104697.

Alpan, K., Tuncal, K., Ozkan, C., Sekeroglu, B., and Ever, Y.K. (2022). *Design and simulation of global model for carbon emission reduction using IoT and artificial intelligence*. Procedia Computer Science 204: 627–634, https://creativecommons.org/licenses/by-nc-nd/4.0.

Arslan, A., Haapanen, L., Hurmelinna-Laukkanen, P., Tarba, S.Y., and Alon, I. (2021). *Climate change, consumer lifestyles and legitimation strategies of sustainability-oriented firms*. European Management Journal. 39: 720–730. DOI 10.1016/j.emj.2021.03.005.

Auer, C., Kriegler, E., Carlsen, H., Kok, K., Pedde, S., Krey, V. et al. (2021). *Climate change scenario services: From science to facilitating action*. One Earth, 4: 20. DOI 10.1016/j.oneear.2021.07.015.

Bakker, K. (2022). *Smart Oceans: Artificial intelligence and marine protected area governance*. Earth System Governance 13: 100141, https://doi.org/10.1016/j.esg.2022.100141.

Bianchini, S., Müller, M., and Pelletier, P. (2022). *Artificial intelligence in science: An emerging general method of invention*. Research Policy, 51, 10, 104604. DOI 10.1016/j.respol.2022.104604.

Bloomfield, P.S., Clutton-Brock, P., Pencheon, E., Magnusson, J., Karpathakis, K. (2021). *Artificial Intelligence in the NHS: Climate and Emissions*. The Journal of Climate Change and Health 4, 100056, https://doi.org/10.1016/j.joclim.2021.100056.

Campbell, J., Jensen, D., Kim, A., and Theresa, D. (2019). Building a Digital Ecosystem for the Planet, Foresight Brief, 14.

Chakraborty, D., Alam, A., Chaudhuri, S., Başağaoğlu, H., Sulbaran, T., and Langar, S., (2021). *Scenario-based prediction of climate change impacts on building cooling energy consumption with explainable artificial intelligence*. Applied Energy 291: 116807, https://doi.org/10.1016/j.apenergy.2021.116807.

Chester, K.M., and To K.P. Chau Chi Wai Kan. (2019). *The logic of innovative value proposition: A schema for characterizing and predicting business model evolution*. Journal of Business Research.

Chung, W.J., and Liu, C. (2022). *Analysis of input parameters for deep learning-based load prediction for office buildings in different climate zones using eXplainable Artificial Intelligence*. Energy & Buildings 276: 112521, https://doi.org/10.1016/j.enbuild.2022.112521.

Cukier, K. (2019). *Ready for robots: how to think about the future of ai*, Foreign Affairs, 98: 192.

Cully, A., Clune, J., Tarapore, D., and Mouret, J.-.B. (2015), *Robots that can adapt like animals*, Nature, 521, 7553, pp. 503–507.

Das, K.P., and Chandra, J. (2023). *A survey on artificial intelligence for reducing the climate footprint in healthcare*, Energy Nexus 9 100167, https://doi.org/10.1016/j.nexus.2022.100167.

de Abreu, M.C.S., Webb, K., Araújo, F.S.M., and Cavalcante, J.P.L. (2021). *From "business as usual" to tackling climate change: Exploring factors affecting low-carbon decision-making in the Canadian oil and gas sector*. Energy Policy, 148, A, 111932. DOI 10.1016/j.enpol.2020.111932.

De Felice, F., Petrillo, A., De Luca, C., and Baffo I. (2022). *Artificial Intelligence or Augmented Intelligence? Impact on our lives, rights and ethics*. Procedia Computer Science, 200: 1846–1856, DOI 10.1016/j.procs.2022.01.385.

de Lamotte, M. (2020). *Enlightenment, Artificial Intelligence and Society*. IFAC PapersOnLine, 53, 2, pp. 17427–17432. DOI 10.1016/j.ifacol.2020.12.2110.

Demertzidis, N., Tsalis, T.A., Loupa, G., and Nikolaou, I.E. (2015). *A benchmarking framework to evaluate business climate change risks: A practical tool suitable for investors decision-making process*. Climate Risk Management, 10, s. 95–105. DOI 10.1016/j.crm.2015.09.002.

Entezari, A., Aslani, A., Zahedi, R., and Noorollahi, Y. (2023). *Artificial intelligence and machine learning in energy systems: A bibliographic perspective*. Energy Strategy Reviews 45: 101017, https://doi.org/10.1016/j.esr.2022.101017.

Erlandsson, R., Bjerke, J.W., Finne, E.A., Myneni, R.B., Piao, S., Wang, X. et al. (2022). *Remote Sensing of Environment* 280: 113201, https://doi.org/10.1016/j.rse.2022.113201.

Farayola, O.A., Abdul, A.A., Irabor, B.O., and Okeleke, E.C. (2023). *Innovative Business Models Driven By Ai Technologies: A Review*, Computer Science & IT Research Journal P-ISSN: 2709-0043, E-ISSN: 2709-0051 Volume 4, Issue 2, P.85-110, DOI: 10.51594/csitrj.v4i2.608.

Foster, L., Szilagyi, K., Wairegi A., Oguamanam, C., and de Beer J. (2023). *Smart farming and artificial intelligence in East Africa: Addressing indigeneity, plants, and gender*. Smart Agricultural Technology 3: 100132, https://doi.org/10.1016/j.atech.2022.100132.

Francisco, M. (2023). *Artificial intelligence for environmental security: national, international, human and ecological perspectives*. Current Opinion in Environmental Sustainability", 61, 101250. DOI 10.1016/j.cosust.2022.101250.

Galaz, V., Centeno, M.A., Callahan, P.W., Causevic, A., Patterson, T., Brass, I. et al. (2021). *Artificial intelligence, systemic risks, and sustainability*. Technology in Society 67: 101741, https://doi.org/10.1016/j.techsoc.2021.101741.

Galaz, V., Centeno, M.A., Callahan, P.W., Causevic, A., Patterson, T., Brass I. et al. (2021). Artificial intelligence, systemic risks, and sustainability. Technology in Society, 67: 101741. DOI 10.1016/j.techsoc.2021.101741.

Gómez-Orellana, A. M. Guijo-Rubio, D., Pérez-Aracil, J., Gutiérrez, P. A., Salcedo-Sanz, S., Hervás-Martínez, C. (2022). *One month in advance prediction of air temperature from Reanalysis data with eXplainable Artificial Intelligence techniques*. Atmospheric Research, https://doi.org/10.1016/j.atmosres.2023.106608.

Harfouche, A. L., Jacobson, D. A., Kainer, D., Romero, J. C., Harfouche, A.H., Scarascia Mugnozza, G. et al. (2019). *Accelerating Climate Resilient Plant Breeding by Applying Next-Generation Artificial Intelligence*. Trends in Biotechnology, 37(11). https://doi.org/10.1016/j.tibtech.2019.05.007.

Hewitt, C.D., and Stone, R. (2021). *Climate services for managing societal risks and opportunities*. Climate Services, 23: 100240, DOI 10.1016/j.cliser.2021.100240.

Houghton J. (2011). *Global warming, climate change and sustainability*. The John Ray Initiative Briefing Paper, 4th edition, 14. https://doi.org/10.1016/j.jbusres.2019.10.023.

Javaid, M., Haleem, A., Haleem Khan, I., and Suman, R. (2022). *Understanding the potential applications of Artificial Intelligence in Agriculture Sector*. Advanced Agrochem., https://doi.org/10.1016/j.aac.2022.10.001.

Jenis, J., Ondriga, J., Hrcek, S., Brumercik, F., Cuchor, M., Sadovsky, E. et al. (2023). *Engineering Applications of Artificial Intelligence in Mechanical Design and Optimization. Machines*, 11: 577. https://doi.org/10.3390/machines11060577.

Ji, L., and Huang, X. (2022). *Analysis of social governance in energy-oriented cities based on artificial intelligence*, Energy Reports, 8, s. 11151–11160, DOI 10.1016/j.egyr.2022.08.206.

Jiang, S., and Li, J. (2020). *Global climate governance in the new era: Potential of business actors and technological innovation*. Chinese Journal of Population, Resources and Environment, 18(3): 165–171. DOI 10.1142/S234.5748.121500081.

Karanth, S., Benefo, E.O., Patra, D., and Pradhan, A.K. (2023). *Importance of artificial intelligence in evaluating climate change and food safety risk*. Journal of Agriculture and Food Research, 11: 100485, DOI 10.1016/j.jafr.2022.100485.

Klaus, G., Ernst, A., and Oswald, L. (2020). *Psychological factors influencing laypersons' acceptance of climate engineering, climate change mitigation and business as usual scenarios*. Technology in Society, 60: 101222. DOI 10.1016/j.techsoc.2019.101222.

Larosa, F., and Mysiak, J. (2020). *Business models for climate services: An analysis*. Climate Services, 17: 100111. DOI 10.1016/j.cliser.2019.100111.

Lee, M. (2020). *An analysis of the effects of artificial intelligence on electric vehicle technology innovation using patent data*. World Patent Information, https://doi.org/10.1016/j.wpi.2020.102002.

Liu, Z., Gao, Y., and Liu, B. (2022). *An artificial intelligence-based electric multiple units using a smart power grid system*, Energy Reports 8: 13376–13388, https://doi.org/10.1016/j.egyr.2022.09.138.

Maghami, M.R., and Mutambara, A. (2023). *Challenges associated with Hybrid Energy Systems: An artificial intelligence solution*, Energy Reports 9: 924–940, https://doi.org/10.1016/j.egyr.2022.11.195.

Mariani, M.M., Machado, I., and Nambisan, S. (2023). *Types of innovation and artificial intelligence: A systematic quantitative literature review and research agenda*. Journal of Business Research, 155: 113364. DOI 10.1016/j.jbusres.2022.113364.

Matos, S., Viardot, E., Sovacool, B.K., Geels, F.W., and Xiong, Y. (2022). Innovation and climate change: A review and introduction to the special issue. Technovation, 117: 102612. DOI 10.1016/j.technovation.2022.102612.

McCarthy, J., Minsky, M. L., Rochester, N., and Shannon C. E. (2006). *A Proposal for the Dartmouth Summer Research Project on Artificial Intelligence, August 31, 1955*. AI Magazine 27(4): 12–14.

Mishra, S., Tripathi, A. R. (2021). *AI business model: an integrative business approach*. Journal of Innovation and Entrepreneurship, 10: 18, https://doi.org/10.1186/s13731-021-00157-5.

Nasir, O., Javed, R.T., Gupta, S., Vinuesa, R., and Qadir, J. (2023). *Artificial intelligence and sustainable development goals nexus via four vantage points*. Technology in Society, 72: 102171. DOI 10.1016/j.techsoc.2022.102171.

Negnevitsky, M. (2011). *Artificial Intelligence: A Guide to Intelligent Systems*, Addison Wesley, Harlow.

Neubauer, A.C. (2021). *The future of intelligence research in the coming age of artificial intelligence – With a special consideration of the philosophical movements of trans- and posthumanism*. Intelligence, 87: 101563. DOI 10.1016/j.intell.2021.101563.

Nti, E.K., Cobbina, S.J., Attafuah, E.E., Opoku, E., and Gyan, M.A. (2022). *Environmental sustainability technologies in biodiversity, energy, transportation and water management using artificial intelligence: A systematic review*. Sustainable Futures 4: 100068, https://doi.org/10.1016/j.sftr.2022.100068.

Pandyaswargo, A.H., Maghfiroh, M.F.N., and Onoda, H. (2023). *Global distribution and readiness status of artificial intelligence application on mobility projects*. Energy Reports 9: 720–727, https://doi.org/10.1016/j.egyr.2022.11.070.

Paul, A., Lang, J.W.B., and Baumgartner, R.J. (2017). *A multilevel approach for assessing business strategies on climate change*. Journal of Cleaner Production, 160(1): 50–70, DOI 10.1016/j.jclepro.2017.04.030.

Polas, M. R. H., Afshar Jahanshahi, A., Kabir, A. I., Sohel-Uz-Zaman, A. S. M., Osman, A. R., and Karim, R. (2022). *Artificial Intelligence, Blockchain Technology, and Risk-Taking Behavior in the 4.0IR Metaverse Era: Evidence from Bangladesh-Based SMEs*. Journal of Open Innovation: Technology, Market, and Complexity, 8: 3. DOI 10.3390/joitmc8030168.

Reddy, S., Fox, J., and Purohit, M.P. (2019). *Artificial intelligence-enabled healthcare delivery*. Journal of the Royal Society of Medicine, 112, 1, s. 22–28. DOI 10.1177/0141076818815510.

Reim, W., Åström, J., and Eriksson, O. (2020). Implementation of Artificial Intelligence (AI): A Roadmap for Business Model Innovation, AI, 1, 180–191; doi:10.3390/ai1020011.

Reine Nishimwe, A.M., and Reiter, S. (2022). *Using artificial intelligence models and degree-days method to estimate the heat consumption evolution of a building stock until 2050: A case study in a temperate climate of the Northern part of Europe*. Cleaner and Responsible Consumption 5: 100069, https://doi.org/10.1016/j.clrc.2022.100069.

Rochlin, C. (2021). *Climate change and the progressive business model*. The Electricity Journal, 34: 106927. DOI 10.1016/j.tej.2021.106927.

Rosenstock, T.S., Lubberink, R., Gondwe, S., Manyise, T., and Dentoni, D. (2020). *Inclusive and adaptive business models for climatesmart value creation*. Current Opinion in Environmental Sustainability, 42: 76–81. DOI 10.1016/j.cosust.2019.12.005.

Russell, S.J., and Norvig, P. (2017). *Artificial Intelligence: a Modern Approach*, Prentice Hall, Upper Saddle River, New Jersey.

Ryo, M. (2022). *Explainable artificial intelligence and interpretable machine learning for agricultural data analysis*. Artificial Intelligence in Agriculture 6: 257–265, https://doi.org/10.1016/j.aiia.2022.11.003.

Saheb, T., Dehghani, M., and Saheb, T. (2022). *Artificial intelligence for sustainable energy: A contextual topic modeling and content analysis*. Sustainable Computing: Informatics and Systems, 35: 100699, DOI 10.1016/j.suscom.2022.100699.

Salehnia, N., and Ahn, J. (2022). *Modelling and reconstructing tree ring growth index with climate variables through artificial intelligence and statistical methods*. Ecological Indicators 134: 108496, https://doi.org/10.1016/j.ecolind.2021.108496.

Sánchez, J.M., Rodríguez, J.P., and Espitia, H.E. (2022). *Bibliometric analysis of publications discussing the use of the artificial intelligence technique agent-based models in sustainable agriculture*. Heliyon, 8: e12005, https://doi.org/10.1016/j.heliyon.2022.e12005.

Schalkoff R.J. (1990). *Artificial Intelligence: an Engineering Approach*, McGraw-Hill, New York.

Sigüenza, C.P., Cucurachi, S., and Tukker, A. (2021). *Circular business models of washing machines in the Netherlands: Material and climate change implications toward 2050*. Sustainable Production and Consumption, 26: 1084–1098. DOI 10.1016/j.spc.2021.01.011.

Sivapuram, V.R.K.P., and Shaw R. (2020). *International investments and businesses as enablers of globalization of local risks: A case for risk communication and climate fragility reduction*. Progress in Disaster Science, 8: 100125. DOI 10.1016/j.pdisas.2020.100125.

Strelkova, O., and Pasichnyk, O. (2017). *Three Types of Artificial Intelligence*. Khmelnitsky National University, Khmelnytskyi, Ukraine, pp. 1–4.

Subeesh, A., and Mehta, C.R. (2021). *Automation and digitization of agriculture using artificial intelligence and internet of things*. Artificial Intelligence in Agriculture 5: 278–291, https://doi.org/10.1016/j.aiia.2021.11.004.

Sugar, K. (2019). *Ready for robots: how to think about the future of ai*. Foreign Affairs, 98: 192.

Swim, J.K., and Geige, N. (2021). *Policy attributes, perceived impacts, and climate change policy preferences*. Journal of Environmental Psychology, 77: 101673, DOI 10.1016/j.jenvp.2021.101673.

Taddeo, M., Tsamados, A., Cowls, J., and Floridi, L. (2021). *Artificial intelligence and the climate emergency: Opportunities, challenges, and recommendations*. One Earth, 4, 6, pp. 776–779. DOI 10.1016/j.oneear.2021.05.018.

Tadesse, D. (2010). *The impact of climate change in Africa*, "Institute for Security Studies Papers, 220, ISSN 1026-0404.

Tariq, R., Ali, M., Sheikh, N.A., Shahzad, M.W., and Xu, B.B. (2023). *Deep learning artificial intelligence framework for sustainable desiccant air conditioning system: Optimization towards reduction in water footprints*, International Communications in Heat and Mass Transfer 140: 106538, https://doi.org/10.1016/j.icheatmasstransfer.2022.106538.

The Royal Society (2009). *Geoengineering the Climate: Science Governance and Uncertainty*, The Royal Society, London, UK.

Tukker, A. (2004). *Eight types of product–service system: eight ways to sustainability? Experiences from SusProNet*. Business Strategy and the Environment, 13, 4, pp. 246–260. DOI 10.1002/bse.414.

Wainaina, S., and Taherzadeh, M.J. (2023). *Automation and artificial intelligence in filamentous fungi-based bioprocesses: A review, Bioresource Technology* 369: 128421, https://doi.org/10.1016/j.biortech.2022.128421.

Wan, J., Li, X., Dai, H.N., Kusiak, A., Martinez-Garcia, M., and Li, D. (2020). *Artificial intelligence-driven customized manufacturing factory: key technologies, applications, and challenges. Proceedings of the IEEE*, 109(4): 377–398. doi: 10.1109/JPROC.2020.3034808.

Wang, C. (2021). *The role of technological innovation in global climate policy*. Research in Globalization, 3: 100066. DOI 10.1016/j.resglo.2021.100066.

Weber, M., Beutter, M., Krcmar, H., Weking, J., and Bohm, M. (2022). *AI Startup Business Models, Key Characteristics and Directions for Entrepreneurship Research*. Bus. Inf. Syst. Eng. 64(1): 91–109, https://doi.org/10.1007/s12599-021-00732-w.

Xiang, Y., Chen, Y., Xu, J., and Chen, Z. (2022). *Research on sustainability evaluation of green building engineering based on artificial intelligence and energy consumption*, Energy Reports 8: 11378–11391, https://doi.org/10.1016/j.egyr.2022.08.266.

Xu, Y., Xingping, Z., Huihui, L., Zheng, H., Zhang, J., Olsen, M.S. et al. (2022). *Smart breeding driven by big data, artificial intelligence, and integrated genomic-enviromic prediction.* Molecular Plant, 15: 1664–1695, https://doi.org/10.1016/j.molp.2022.09.001.

Zhang, L.a., Ling, J.b., and Lin, M. (2022). Artificial intelligence in renewable energy: A comprehensive bibliometric analysis. Energy Reports 8: 14072–14088, https://doi.org/10.1016/j.egyr.2022.10.347.

Zhou, Y. (2022). *Artificial intelligence in renewable systems for transformation towards intelligent buildings*, Energy and AI 10 100182, https://doi.org/10.1016/j.egyai.2022.100182.

Chapter 6

Climate Hazard Management Mechanisms using Artificial Intelligence Solutions

6.1 Introduction

The purpose of this chapter is to present mechanisms for managing climate hazards using artificial intelligence solutions.

The scope of the chapter covers the issues of climate security and cybersecurity and the package of artificial intelligence solutions for managing climate hazards.

The chapter deals with the place and role of artificial intelligence in the processes of effective and efficient management of climate hazards.

6.2 Artificial intelligence in climate hazard management

Climate change is often examined too narrowly, focusing on the natural aspect in the analysis thereof without a multidimensional look at other areas of theoretical and practical influences. A factor broadening the cognitive perspectives in this topic is appropriate risk management mechanisms in many dimensions. Appropriate risk management mechanisms are a platform for the development of strategic and tactical–operational control of well-managed organizations. Such logic in the face of crisis situations, which are unfortunately now the natural working environment of companies, must be adopted and applied. The survival of many types of organizations depends on it, and failure to do so can be very dangerous. Integrated risk management is the foundation of modern business management from the perspective of real business associated with a number of risks inherent in a difficult socio-economic ecosystem. The concept is embedded in the logic of understanding risk.

DOI: 10.1201/9781003434016-6

Many definitions and types of risk occur. The proper understanding of the potential inherent in integrated risk management mechanisms allows for the skillful avoidance of hazards or their appropriate mitigation. Mitigation in this case is understood as weakening these hazards or eliminating them completely.

It is very important to define the types of risk and their proper interpretation from the perspective of risk management as one of the key management concepts at present. Usually, in organizations of various types, we focus on explicit risk, which is a risk that we can imagine and try to materialize over many years to the best of the state of current knowledge and historical conditions. Such risks can be managed with awareness of the risks arising therefrom. On the other hand, implicit risk cannot be managed because it cannot be imagined. Internal risk can be managed, as this risk results directly from the resources within the organization. External risks cannot be managed. Instead, there can be a strong focus on the appropriate response to it and its impact on the company. Hence, an adequate and appropriate response to this risk becomes crucial. This risk is located in the socio-economic ecosystem where the company conducts its business activities. Finally, the risk can be active and passive. The former is related to the decision-making processes carried out by the managers of a given organization and the latter is our response to the decision-making processes of managers of companies competing with a given company. Risks can be hierarchical and then we face risks at strategic, tactical and operational level. Risk can also have a functional dimension. Such risks may include, for example, purchasing, marketing, production, investment, personnel, warehouse risks, etc. Proper qualification and quantification of risk is a crucial determinant of the effectiveness and efficiency of an organization from the perspective of building integrated risk management in an enterprise. Looking at the organization in holistic terms, it is necessary to take into account the ignorance of many managers in the context of understanding direct and indirect threats around the organization and, above all, the lack of systemic response to them. An ignorant manager is a significant problem for many organizations and their employees. The lack of multi-criteria analysis and synthesis in risk management can be a dynamic platform for the collapse of even the best companies with the best products and/or services. Hence, a very important element of wise and informed organizational management is to define the place and role of risk management among other management concepts and applicable management paradigms, i.e., solutions, techniques, methods and management concepts sanctioned and recognized by the business and scientific community. From the perspective of the formalization of management systems, which aims to develop mechanisms of repetitiveness of adopted solutions and decision-making processes within companies, it is important to develop a set of norms and standards applicable to a given type of organization, providing a guarantee of logical and systemic supervision over various types of processes, projects, business models and strategies of organizations that are their most important ontological entities determining either the market success of the organization or its failure when the mechanisms of an effective and effective process of managing types of entities are not mastered. This strategic trajectory enables dynamic market progress, ensuring the company's ability to survive in even the most difficult market conditions. Hence, the dynamically emerging environment of many new crises in the economy and society radically changes the functioning mechanisms of many companies around the world. This completely changes the logic of running a business; new, specific conditions create, on the one hand, a number of threats and, on the other hand, open up opportunities and development prospects that so far could not have even been imagined. This is related to the dynamic acceleration of technological and digital development of the economy and the Earth's changing climate, which results in the current stream of many hazards, which should be comprehensively counteracted. To survive, businesses must manage business continuity by reacting systematically to all positive and negative signals, adapting to any changes through modifications to their business models and strategies, including crisis strategies. Nowadays, examples of significantly overlapping crises are as follows:

1. The economic and social crises triggered, among others, by the dynamic rise in inflation, rising energy costs and credit service costs at the level of companies and individuals,
2. The political crisis at the national and European level expressed by clashing and very rapidly changing approaches to business, the economy and society, polarizing the behavior of social groups and individuals,
3. The health crisis caused by the COVID-19 pandemic undermining the well-being of society as a whole,
4. The military and migration crisis caused by the war in Ukraine, which provokes opportunistic behavior among European societies and changes the structure of employment on the labor market,
5. The technology crisis caused by changes related to the dynamic development of new technologies, automation and robotics, which results, among others, in negative situations on the labor market eliminating the performance of simple operational activities by people in favor of machines, robots and industrial automation mechanisms and digital exclusion of an ageing society in Poland and Europe,
6. The moral–ethical crisis related to the dynamic change of values at the level of various social groups and individuals.

In such a cognitive space, where business, the economy and society exist, a widespread multidimensional and multi-criteria stream of existing pressures, limitations, threats, and opportunities occurs.

The ongoing digital and climate transformations have a significant impact on the socio-economic phenomena taking place at the same time. Digital transformation creates new organizational and human behaviors using a virtual ecosystem embedded in the Internet, where various groups of the digital community function and meet their needs (so-called netizens).

In turn, climate transformation is a strong response to dynamic climate changes, the effects of which materialize in the form of many physical threats of varying intensity that adversely affect the entire planet Earth and its vital and existential functions. In such spatial conditions, companies and communities search for the optimal solutions of various natures, which will be the determinant of their survival even in the most difficult conditions. These optimal solutions are placed in the economy and society and their recipients are wide groups of actors forming a system of interrelated relationships, which are mutually dependent in many respects. In this dimension, companies create their new, unique, sophisticated business models and strategies based on social, technological and digital innovations. From this perspective, dynamic managers pursue their strategic and tactical–operational goals using the technological and digital potential currently being developed, fulfilling functions relevant to high performance organizations (HPOs) and high reliability organizations (HROs).

AI solutions have come to the forefront. Artificial intelligence is treated as a mapping of people's learning processes, a new dimension of processing thought mechanisms and shaping new platforms for generating intelligence. These solutions are aimed, among others, at creating new and individualized business models supported by the development of new economies, for example, the sharing economy, the circular economy and the network economy, the Big Data economy, etc. Artificial intelligence and the inherent potential therein goes to develop the streams and spaces of emerging skill gaps in the direction of identifying crucial problems for business, the economy and society, and creating paths towards solving them. Such a relationship occurs, among others, when solving climate problems in Europe and beyond. It can be solved by the power and potential of artificial intelligence, mainly in the context of weakening vulnerability to climate hazards and their impact on all spheres of human existence on earth. Therefore, appropriate logic and methodology for exerting a positive impact on the natural environment and the dynamic, uncontrolled phenomena of climate change can be sought by means of precisely dedicated and targeted principles of artificial intelligence. In this case, it can be implemented by the following:

1. A concrete, precise and innovative approach to technical, organizational–management and tactical–operational solutions in the field of combined digital and climate transformation.
2. Designing and applying unique digital business models and strategies based on developed artificial intelligence algorithms.
3. Using artificial intelligence to technologically manage companies changing their operation logic to positively impact climate change.
4. Linking AI solutions to the creation of new services to evoke positive climate change and the implementation of the principles determining the climate neutrality of the economy.
5. Creating new AI-based technical enterprises to mitigate climate hazards in conjunction with the use of renewable energy resources as a primary source for the production of their products and services.
6. Digital transformation as a new dimension of shaping reality supported by innovative solutions increasing the energy efficiency of the economy.
7. Climate transformation as a dynamic accelerator of socio-economic changes supported by the implemented principles of impact investing.
8. Artificial intelligence as an important factor in the development of modern business and social ecosystems for the joint achievement of high social and economic efficiency of enterprises.
9. Virtual reality as an accelerator for creating new digital business models and strategies compared with analog solutions based on direct market contact.
10. Automation and robotics in the processes of shaping virtual reality, taking into account the human factor and humanoids.
11. Digital business models and strategies as a platform for creating a positive impact on the environment, taking into account the assumptions of the circular economy.
12. Developing digital platforms based on algorithms and Big Data.
13. Creating new climate services by applying new economies to produce a positive impact on the climate.
14. Creating local electric power ecosystems (energy clusters) ensuring energy self-sufficiency of regions through the efficient use of renewable energy sources (the use of wind, sun, water, bio-waste for electricity, heat, gas, etc.)

15. Developing creative AI-enabled solutions related to the technical, physical and energy security of a country's critical infrastructure facilities.
16. Technical and terrorist cybersecurity provided through solutions contained in artificial intelligence algorithms and in the digital business models of companies with quantum data processing processes used therein.
17. The practical implications of using artificial intelligence for positive climate change.

In such a defined stream of decision-making and implementation of solutions, a modern and original approach to the design and implementation of effective and efficient AI solutions ensuring the induction of positive climate change is determined. Then, the use of digital business models and strategies combining the solutions of digital and climate transformations provides an important stimulus and a driver for evoking positive climate change. The presented solutions, including digital platforms based on algorithms, significantly change the functioning of socio-technological organizations. Moreover, the applied principles of technological management of companies strengthened by the elements of automation and robotics will significantly and clearly shape the emergence of new, positive climate services, and the mechanisms for implementing the principles of cybersecurity and security at the level of companies will serve to dynamically reduce current climate hazards. It should also be noted that a large amount of AI research into climate change focuses on forecasting energy demand (Ahmad et al. 2019, pp. 242–250). This refers to the use of predictive models in forecasting mechanisms, which can be divided into the following types (Ahmad and Chen 2018, pp. 410–417): physically oriented, data-oriented, hybrid and large-scale.

6.3 The largest sources of uncertainty related to global warming and climate change and the effects of climate change on human and animal health versus artificial intelligence solutions

In a world full of various types of threats, we function in a wide range of uncertainties, risks and certainty and a network of paradigms and dogmas. The core sources of uncertainty related to global warming and climate change should be addressed in a systemic manner. The mechanisms of problem management concepts can be used for this purpose. Problem management is conditioned by the result of the analysis of data from the incident identification and management process carried out by a trained and dedicated analyst. On this basis, he identifies key incidents that will be subject to appropriate further root cause analysis (RCA) (Wilson et al. 1993, pp. 8–17). Climate problem management reveals, among others, the effects of climate change on humans and animals. For example, rising temperatures, changes in ecosystems and extreme weather phenomena lead to a deterioration in air quality, the spread of infectious diseases and increased heat- and stress-related health problems.

Climate change has a number of adverse health effects, often seemingly unrelated to the temperature rise itself. The WHO estimates that climate change is directly responsible for more than 140,000 deaths a year, primarily in Africa and Southeast Asia (Harris et al. 2017). Table 13 below shows the impact of climate change on human health.

As the table shows, the impact of climate change on human health results from different types of climate hazards. They concern both phenomena occurring continuously and those which are sudden and often uncontrollable with a direct impact. It is therefore necessary to have a different set of responses to this, in both predictive and reactive terms.

6.4 National, regional and global security and cybersecurity and the impact of climate change

Security and cybersecurity are extremely important for the implementation of business processes based on digital business models supported by artificial intelligence solutions. The digitalization of business models introduces new methods, techniques and tools for doing business. For example, the principles of business monetization and the implementation of activities using micropayments by means of digital business models are clear evidence of the automation of commercial transactions supported by modern revenue generation systems based on digital payment models.

Security and cybersecurity aspects are currently shaping most decision-making processes around the world. Safety, understood as a

Table 13. Climate change impact on human health.

No.	Climate change impact on human health
1.	Heat waves
2.	Water shortages
3.	Extreme weather phenomena
4.	Infectious diseases
5.	Air pollution
6.	Mental health
7.	Migrations

Source: Own study.

subjective condition in which hazards are under control, is very difficult to obtain. Therefore, it must be properly and wisely managed. This requires specialist knowledge, which not only must be possessed but above all must be used by managers in their management processes. It is not so simple and does not always bring satisfactory results in the short and long term. This is because the subject is very complex, especially with regard to external conditions affecting various types of companies in a dangerous way.

When examining safety and cybersecurity aspects, it is crucial to have a clear understanding of the concepts of safety, cybersecurity, physical protection and cyber protection.

Table 14 below explains the core safety, digital safety, security and cybersecurity concepts.

It should be noted that, in accordance with Regulation (EU) 2025/38 of the European Parliament and of the Council of 19 December 2024 establishing measures to strengthen solidarity and capacities in the European Union to detect, prepare for and respond to cyber threats and incidents, and amending Regulation (EU) 2021/694 (Cyber Solidarity Act):

- At European Union and global levels, the scale, frequency and impact of cybersecurity incidents are increasing, including cyber-espionage attacks against supply chains, the installation of ransomware or disruptions. They pose a serious threat to the functioning of network and information systems. In view of the rapidly changing threat landscape, the threat of possible large-scale cybersecurity incidents causing a serious disruption or damage to critical infrastructure

Table 14. Clarification of key safety and security concepts.

No.	Concept	Explanation of the Concept
1.	Safety	A condition in which analog threats to the company's operational activities are under control. It concerns the functioning of processes, projects, business models and strategies in internal terms.
2.	Digital Safety	A state in which cyber threats related to the company's operational activities are under control. It concerns the functioning of processes, projects, business models and strategies in internal terms.
3.	Security	A state in which the company's internal resources, including its critical infrastructure, are protected through appropriate physical protection mechanisms.
4.	Cybersecurity	A state in which the company's internal resources, including its critical infrastructure, are protected through appropriate physical cyber protection mechanisms.

Source: Own study.

requires the increased preparedness of the European Union cybersecurity framework.

- A European Cybersecurity Alert System, a pan-European infrastructure network consisting of national cybersecurity hubs and cross-border cybersecurity hubs, joining on a voluntary basis, is established to support the development of advanced European Union capabilities for the detection, analysis and processing of data in relation to cyber threats and incident prevention in the European Union.
- A Cyber Emergency Mechanism is established to support the improvement of the European Union's resilience to cyber threats and to prepare for, and mitigate in a spirit of solidarity, the short-term impact of significant cybersecurity incidents, large-scale cybersecurity incidents and incidents equivalent to large-scale cybersecurity incidents.

In order to build an appropriate approach to safety and cybersecurity management, it is important to apply scenario planning solutions and principles in the area of strategic climate change management. It should be noted that scenario planning is based on defining appropriate strategic options. These options may be subdivided into a minimum, averaged and maximum plan or targeted towards so-called options A, B, C or D.

It is also important to be aware that AI solutions used in a hostile way can create disinformation mechanisms that sometimes even cause complete disorientation of societies in the area of real scientific knowledge of climate change. At the same time, artificial intelligence in itself can generate a number of threats with different strengths of economic and social impact.

Table 15 below provides a list of selected AI-related hazards.

Table 15. List of selected AI-related hazards.

No.	Selected AI-related hazards
1.	Job losses linked to the automation of occupations
2.	Transforming social relationships
3.	Misuse of AI by governments
4.	Use of AI for disinformation and manipulation
5.	Loss of privacy
6.	AI as a weapon
7.	Uncontrolled development of AI

Source: Own study.

These hazards can occur individually, but also together. Then, the degree of danger increases disproportionately and often very uncontrollably. Controlling such a stream of hazards then requires extremely sophisticated neutralization strategies. In such cases, it is necessary to build tight systems for managing relevant climate security criteria after describing them and determining their potential impact on socio-technical ecosystems.

6.5 Climate security and artificial intelligence

Security is treated as a subjective state in which all hazards are under control. Climate security can be considered one of its sub-categories. The problem, however, is in the categorization of this particular sub-category. It can be classified as national, procedural, technological or social security. In each of these references, it will have a different definition and a different logical interpretation. The discussion on national climate security highlights military applications and the role of artificial intelligence in disinformation and climate denial. The discussion on international security suggests that international organizations can use artificial intelligence to fulfil their mission and secure their position among international actors. Transnational companies can also benefit from artificial intelligence, which can have an adverse impact on the environment. Artificial intelligence can additionally provide transnational companies and states with tools to obstruct environmental movements. The discussion on human safety highlights the role of artificial intelligence in achieving SDGs, but it can ignore potential power imbalances. Finally, green climate security highlights the role of algorithms in shaping our vision of the environment and how it potentially distances us from other worldviews and the impact of artificial intelligence on the environment (Francis 2023). If underused or misused, artificial intelligence can undermine current environmental policies, slow down efforts to support sustainable development and impose severe environmental costs on current and future generations. However, if used properly, artificial intelligence can be a powerful tool to develop effective responses to the climate crisis. Policymakers and the scientific community must act urgently to ensure that this impact is as positive as possible, in the interest of a fair and sustainable future (Taddeo et al. 2021). Various stakeholders exhibit unsustainable behavior. Such behavior comes from reckless consumers, producers of unsustainable energy sources and decision-makers who cannot govern and control these unsustainable behaviors. Machine learning algorithms and autonomous robots are examples of artificial intelligence in its soft and hard forms that can be used to monitor, manage and direct the behavior of various stakeholders (Saheb et al. 2022). Food security is one of the priorities of every country in the world. However, various factors hinder the achievement of global food security objectives. Some unprecedented shocks are hampering food security at the global level. Different interventions have been used to ensure food security, and artificial intelligence is one of the modern methods used at different stages of the food system. The study assessed the use of artificial intelligence across the entire food production ecosystem, including crop production, animal production, harvest/slaughter, post-harvest management, food processing, food distribution, food consumption and food waste management. The aim of this study was to assess the use of AI systems at all stages of food systems. A systematic review was carried out, analyzing 110 articles after searching 450 articles on the basis of inclusion and exclusion criteria. The results indicated that different AI algorithms are used at all stages of the food system, from plant/animal production to food or agricultural waste management (Kutyauripo et al. 2023). Due to the improper management of food products at various stages of production and processing, the problem of hunger and malnutrition is persistent.

The introduction of modern technologies is a boon in response to the devastation of malnutrition and food security. Metagenomic research can ensure food security in terms of pathogens or diseases for food products, as well as longevity. On the other hand, artificial intelligence helps to determine the demand and usability of a product among consumers. AI components such as e-nose, e-language, e-eye and sorting devices are helpful in reducing food waste and food poisoning. This artificial intelligence venture leads to minimal waste of food resources (Sharma et al. 2022). Climate security criteria have an impact on numerous sectors of the economy, thus triggering so-called decarbonization pressure, which causes dynamic technological changes in many sectors and businesses embedded therein. An example of such a phenomenon is the dynamic technological changes in the civil aviation sector. This applies, among others, to the modernization of aircraft and the implementation of sustainable technological solutions in the broad sense of the word.

6.6 Critical infrastructure management by means of artificial intelligence versus climate change

Critical infrastructure for the security of a country or region is an important strategic resource, which should be exceptionally protected, including against anomalies and climate hazards. The concept of artificial intelligence can be used to supervise and monitor it. Critical infrastructure means an asset, system or part thereof located in Member States which is essential for the maintenance of vital societal functions, health, safety, security, the economic or social well-being of people, and the disruption

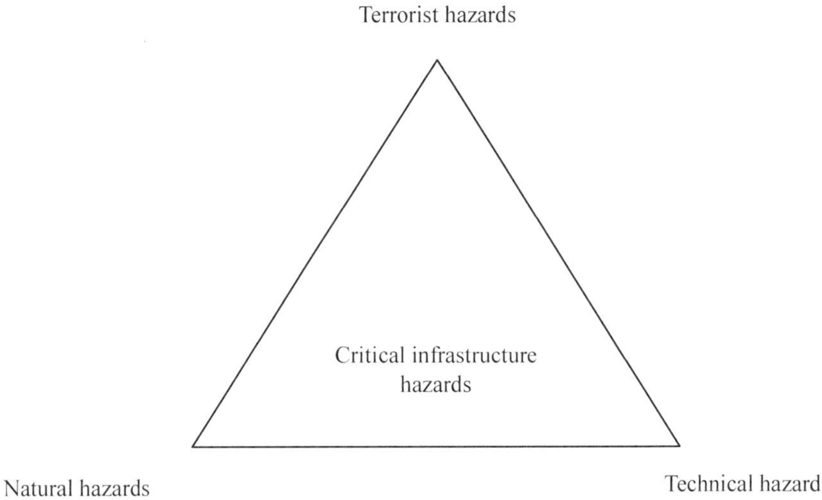

Figure 19. Hazard triangle for critical infrastructure. Source: Own study.

or destruction of which would have a significant impact in a Member State as a result of the failure to maintain those functions (Council Directive 2008/114/EC).

The issue of protection and management of critical infrastructure fits into a wide range of issues related to the concept of crisis management. There is a whole set of hazards that can affect critical infrastructure, from natural and technical hazards to terrorist threats. Therefore, effective and efficient management of critical infrastructure should include the use of state-of-the-art solutions, including AI solutions. Figure 19 shows the hazard triangle for critical infrastructure.

This hazard triad related to critical infrastructure combines the relationships and links related to terrorist, natural and technical threats into one coherent system. Specific mitigation and neutralization actions are dedicated to each group of these hazards. Their systemic approach and consideration generates a broad range of decision-making processes with a strong impact on climate hazards and beyond.

6.7 Artificial intelligence mechanisms in climate hazard management, related constructs

AI mechanisms in climate hazard management can and should be considered specific and functional interlinked constructs. Then, the relationship will be expressed in appropriate conceptual operating, aggregated and dedicated actions. These actions then have the appropriate power of application because they are well-thought-out and relevant to the expected needs in terms of climate action.

Artificial intelligence is increasingly recognized as a powerful and effective tool in the fight against climate change. By using artificial intelligence, huge amounts of environmental, climate and natural data can be analyzed and accurate, detailed climate forecasts can be made based on them. In addition, it allows for the optimization of renewable energy sources and their use and helps obtain data in many other areas as well.

In practice, artificial intelligence is used in terms of climate change for data analysis, predictive modeling and optimization processes.

Data analysis: Artificial intelligence can analyze complex climate data to identify patterns and anomalies that would be difficult for humans to detect.

Predictive modelling: Models can predict climate events, helping us prepare for and mitigate their effects. For example, models simulate climate change based on a set of anthropogenic extortion scenarios, indicating that further greenhouse gas emissions will cause further warming and change in all components of the climate system. Integrated Cost–Benefit Assessment (ICBA) models generate mitigation pathways that maximize prosperity, assuming that deep uncertainties are addressed in future scenarios. These assumptions include socio-economic projections, the scale and dynamics of the climate impact on the economy, and the physical climate response. As the models explore a space of uncertainty within the limits of their objective functions, they risk providing scenarios that are too narrow and not sufficiently robust. In this case, robust and multi-criteria decision-making methods are used to extract relevant information from a large set of optimal emission reduction pathways generated by a regionalized model of integrated cost–benefit assessment with deterministic welfare optimization (Ferrari et al. 2022).

Optimization: Artificial intelligence can optimize energy consumption and reduce waste, helping to reduce the carbon footprint.

6.8 Artificial intelligence, algorithmization and algocracy in climate risk management

Algorithmization is the process of comprehensive and systemic design and development of new algorithms to achieve specific goals and results in a given subject and content area. Algocracy can be understood as the functioning of business, the economy and society in the natural environment of many algorithms. The environment of many algorithms can be treated as a world that creates, processes and functions on the basis of defined and implemented algorithms regulating our living activities on earth. If we want to achieve something, we build the right algorithm by means of artificial intelligence. If we want to change something, we also do it according to the previous scheme. Then, we connect the real world with the virtual one. The link between these two worlds is these algorithms.

Private companies are primarily responsible to shareholders and investors. Companies face increasing pressure from shareholders to address climate hazards (Goldstein et al. 2019, p.18–25). Algocracy can help. Algocracy is a world governed by algorithms, and it is then that they have a huge impact on this world. Such a philosophy is increasingly being implemented in the current space of socio-economic and cyber-technical ecosystems (Rybiński and Królewski 2023). In algocracy, business processes and their criteria required by different stakeholder groups are crucial. Business process models serve a wide range of purposes for which a distinction can be made between models to be read by humans and those to be read by machines (Gschwind et al. 2014, pp. 117–132).

6.9 Artificial intelligence in the climate hazard management audit – security and cybersecurity aspects

An audit is, on the one hand, an active search for compliance and/or a photograph of reality and/or a systematic, independent process to confirm that predetermined requirements are implemented and met. The climate hazard management audit is a very important tool in the processes of assessment and evaluation of adaptation and climate change. Artificial intelligence is widely used in cybersecurity auditing. It supports automation, data analysis, anomaly detection and process optimization, helping to carry out the so-called continuous audit and verify the effectiveness of the implemented control mechanisms (Wygodny 2024). In this case, the climate hazard management audit performs a very important primary function, i.e., that potential vulnerabilities can be identified during the audit in order to determine actions to prevent their occurrence by applying appropriate measures to control these hazards. The research consists of observing climate phenomena, examining documents and talking to key people who can shape climate policy at different levels of the organization or region.

6.10 Weather forecasting and climate change modelling by means of AI solutions

One of the core components of climate change management is the proper forecasting of their changes, with particular emphasis on anomalies, dysfunctions and climate catastrophes, both those caused by human actions and those caused by natural forces. Climate models, which are complex mathematical and computer tools that simulate the processes in the Earth's climate system, are widely used. They are based on the physical movement of the atmosphere, oceans, glaciers and the Earth's surface, as well as on the interactions between them. Artificial intelligence strongly supports the development of new technological mechanisms in meteorology. Thanks to unprecedented computing power, it becomes possible to see even subtle weather patterns. AI models are starting to replace or support current methods, enabling predictions with increased accuracy. Moreover, predicting extreme weather events in the context of climate change is also crucial. By minimizing all types of losses resulting from natural disasters and weather extremes or unforeseen turbulence and weather anomalies, artificial intelligence markedly strengthens the foundations of society and guarantees the stable development of global economies. Therefore, climate modelling and weather prediction needs to be more accurate and effective. At this point, it should be noted that there is a huge lack of understanding of the differences between the weather, which is chaotic and difficult to predict, and the climate, i.e., weather averaged over time.

6.11 Carbon capture, satellite observation and pollution detection by means of artificial intelligence solutions

One of the most interesting and innovative applications of artificial intelligence is the area of carbon capture and storage (CCS). AI algorithms are used to design more efficient systems for capturing carbon dioxide emissions from power plants and other industrial sources. In some cases, artificial intelligence even helps develop new materials that can absorb CO_2 from the air, effectively cleansing our atmosphere of the gases driving climate change.

Satellite observation and pollution detection by means of AI solutions is very important because AI algorithms can analyze images from satellites, detecting anomalies that may indicate environmental and climate pollution or illegal activities in this area.

Adopting and applying appropriate mechanisms to monitor and measure global warming by means of artificial intelligence solutions is a core determinant of effective and efficient climate management. Gases in the lower (tropospheric) atmosphere play a key role in the greenhouse effect, such as water vapor, methane, carbon monoxide (IV) and others – called greenhouse gases. They have the ability to absorb and emit infrared light. Greenhouse gases are characterized by durability and chemical passivity, as indicated by the lengthy duration of their presence in the atmosphere (Kociołek-Balawejder and Stanisławska 2012).

6.12 Carbon footprint verification by means of artificial intelligence

Verification of the carbon footprint plays a particular role in the climate impact monitoring process. Carbon footprint verification can be carried out by means of artificial intelligence solutions. The carbon footprint of a product or service is the assigned amount of greenhouse gas emissions (CO2e) produced at each stage of the life cycle of a product or service, from raw materials, to production, transport, use, to disposal. The Climate Performance Framework (CPF) is one of the key and fundamental indicators included in non-financial reporting related to the ESG concept in accordance with a specific EU Taxonomy. Artificial intelligence enables end-to-end support for the automation of entire supply chains, including manufacturing processes, enabling dedicated robots and machines to quickly perform routine tasks. At the same time, AI technology allows for precise identification of defects and deviations in technological processes, which definitely allows for a rapid response and a large and visible improvement in the quality of production and other supporting processes.

6.13 Climate transformation based on digital solutions as a dynamic process of social and economic change

Transformation can also be called a kind of metamorphosis. A process-based climate metamorphosis will be the expected stream of dynamic changes, including structural ones. Climatology is the study of the environment of our planet. Our planet is a magnificent example of a complex, dynamic system in constant motion that balances energy and sustains a habitable environment. Essentially, climatology is also focused on studying large-scale changes in land, atmosphere, oceans, and the cryosphere over long periods (years, decades, centuries), although some believe shorter (weather) timescales, such as days or weeks, are also part of climate study (Faghmous and Kumar 2014). The climate is currently undergoing a large-scale transformation, also known as the Just Transition. The Just Transition Mechanism is intended to be a crucial tool and a fuse to ensure that the transition to a climate-neutral economy takes place in a fair way, leaving no one behind. The climate transition is strongly linked to the digital transition. The term "twin transition" is used, which refers not only to two simultaneous directions of transformation (green and digital), but also to uniting and intensifying actions that could accelerate the implementation of the necessary changes and bring societies closer to the expected level of transformation (Muench et al. 2022).

The core principles of climate transition applicable to the creation of positive climate change can be defined as follows:

- Building systemic energy security in accordance with the accepted principle of subsidiarity,
- Ongoing investments and organizational solutions must not delay or block the pursuit of full climate neutrality,
- Flexibility in the approach to neutralizing climate rules, in particular on the demand, supply and transmission sides,
- Striving to obtain a 100% share of RES in the applied power mix together with used energy storage and green hydrogen,
- Maximizing the use of digitalization solutions in the development of the energy sector,
- Not causing and eliminating significant adverse impacts on the environment and human health (as a consequence of exposure to different types of emissions and impact of climate change).

The climate transformation can be and is strongly driven by digital solutions.

6.14 Conclusions

The adoption of appropriate and optimal principles for building climate safety should be supported by all possible AI solutions. This is especially important in harnessing the dynamics of anomalies and climate dysfunctions in real time. The logic of smart climate management requires a number of streams of intelligent use of artificial intelligence to mitigate climate hazards. Therefore, strategic decision-makers must have climate awareness and sensitivity based on mindfulness and empathy and strategic reflection related to properly interpreted climate hazards in the processes of human and economic existence.

References

Ahmad, T., and Chen, H. (2018). *Utility Companies Strategy for Short-term Energy Demand for Ecasting Using Machine Learning Based Models*. Sustainable Cities and Society, 39: 401–417. DOI: https://doi.org/10.1016/j.scs.2018.03.002.

Ahmad, T., Chen, H., and Shah, W. (2019). *Effective Bulk Energy Consumption Control and Management for Power Utilities Using Artificial Intelligence Techniques under Conventional and Renewable Energy Resources*. International Journal of Electrical Power & Energy Systems, 109: 242–258. DOI: https://doi.org/10.1016/j.ijepes.2019.02.023.

Council Directive 2008/114/EC of 8 December 2008 on the identification and designation of European critical infrastructures and the assessment of the need to improve their protection.

Faghmous, J. H., and Kumar, V. (2014). A Big Data Guide to Understanding Climate Change: The Case for Theory-Guided Data Science, Big Data September 2014, DOI: 10.1089/big.2014.0026.

Ferrari, L., Carlino, A., Gazzotti, P., Tavoni, M., and Castelletti, A. (2022). *From optimal to robust climate strategies: expanding integrated assessment model ensembles to manage economic, social, and environmental objectives*, Environ. Res. Lett. 17: 084029, https://doi.org/10.1088/1748-9326/ac843b

Francisco, M. (2023). *Artificial intelligence for environmental security: national, international, human and ecological perspectives*, Current Opinion in Environmental Sustainability, 61:101250, https://doi.org/10.1016/j.cosust.2022.101250.

Goldstein, A., Turner, W.R., Gladstone, J., and Hole, D.G. (2019). *The private sector's climate change risk and adaptation blind spots*. Nat. Clim. Change 9, s.18–25.

Gschwind, T., Pinggera, J., Zugal, S., Reijers, H. A., and Weber, B. (2014). A linear time layout algorithm for business process models, Journal of Visual Languages and Computing 25: 117–132, http://dx.doi.org/10.1016/j.jvlc.2013.11.002.

Harris, J.M., Roach, B., and Codur A.-M. (2017). *The Economics of Global Climate Change*, Somerville: Global Development and Environment Institute.

Kociołek-Balawejder, E., and Stanisławska, E. (2012). *Chemia środowiska* [Environmental chemistry]. Uniwersytet Ekonomiczny we Wrocławiu, Wrocław.

Kutyauripo, I., Rushambwa, M., and Chiwazi, L. (2023). *Artificial intelligence applications in the agrifood sectors*, Journal of Agriculture and Food Research 11 100502, https://doi.org/10.1016/j.jafr.2023.100502.

Muench, S., Stoermer, E., Jensen, K., Asikainen, T., Salvi, M. i., and Scapolo, F. (2022). *Towards a green and digital future*. Luxembourg: Publications Office of the European Union. https://doi.org/10.2760/977331.

Rybiński, K., and Królewski, J. (2023). *Algokracja, Jak i dlaczego Sztuczna Inteligencja zmienia wszystko* [Algocracy, How and why artificial intelligence changes everything], PWN, Warszawa.

Saheb, T., Dehghani, M., and Saheb, T. (2022). *Artificial intelligence for sustainable energy: A contextual topic modeling and content analysis*, Sustainable Computing: Informatics and Systems 35 100699, https://doi.org/10.1016/j.suscom.2022.100699.

Sharma, P., Vimal, A., Vishvakarma, R., Kumar, P., porto de Souza Vandenberghe L., Kumar Gaur, V. et al. (2022). *Deciphering the blackbox of omics approaches and artificial intelligence in food waste transformation and mitigation*, International Journal of Food Microbiology 372, 109691, https://doi.org/10.1016/j.ijfoodmicro.2022.109691.

Taddeo, M., Tsamados, A., Cowls, J., and Floridi, L. (2021). *Artificial intelligence and the climate emergency: Opportunities, challenges, and recommendations*, One Earth 4, June 18, https://doi.org/10.1016/j.oneear.2021.05.018

Wilson, P., Dell, L., and Anderson, G. (1993). *Root Cause Analysis: A Tool for Total Quality Management*, Milwaukee, 1993, ASQ Quality Press, s. 8–17.

Wygodny, A. (2024). *Sztuczna inteligencja w audycie i zarządzaniu ryzykiem* [Artificial intelligence in audit and risk management], Kontrola Państwowa, Nr 6, kontrola i audyt DOI: 10.53122/ISSN.0452-5027/2024.1.36.

Chapter 7

Contextualization of the Place and Role of Artificial Intelligence in Climate Hazard Management

7.1 Introduction

The aim of this chapter is to present mechanisms for contextualizing the place and role of artificial intelligence in climate hazard management.

The scope of the chapter covers issues of analog operational security and physical protection systems, as well as operational cybersecurity and cyber protection and climate change along with the definition of secure management of climate data.

The subject of the chapter is the contextualization of the place and the role of artificial intelligence in climate hazard management.

7.2 The state of knowledge of climate hazard management

Currently we do not have full and sufficient knowledge of climate hazards. These include both climatic conditions linked to long-term human intervention in natural ecosystems and uncontrolled climatic events and catastrophes, which are caused by natural factors such as earthquakes and volcanic eruptions. Climate risk is therefore very difficult to determine, a state of affairs which, unfortunately, climate sceptics or political populists are using to achieve short-term public benefits. The state of knowledge of climate hazard management is related to the adopted principles of crisis management in the area of dysfunctions, anomalies, extremes, incidents, accidents and climate catastrophes.

Table 16 below shows a selection of direct or indirect climate hazards.

DOI: 10.1201/9781003434016-7

Table 16. Selected direct or indirect climate hazards.

No.	Selected direct or indirect climate hazards
1.	Epidemic
2.	Flood
3.	Disruption of ICT systems and networks
4.	Drought/heat
5.	Epizootic
6.	Disruption of the energy system
7.	Strong wind
8.	Disruption of the fuel system
9.	Disruption of telecommunications systems and services
10.	Chemical contamination on land and sea
11.	Disruption of the gas system
12.	Railway, air, maritime disaster
13.	Terrorist incident
14.	Radioactive contamination
15.	Large area fire
16.	Epiphytotic
17.	Heavy frost/heavy snowfall

Source: Own study.

This table clearly shows the huge extent of these hazards and, furthermore, the potentially catastrophic adverse effects resulting therefrom. This can neither be underestimated nor trivialized, which unfortunately takes place in various political and social fora. A skillful study of these hazards can offer a great deal of strategic information towards the implementation of effective corrective, preventive and improvement actions in this core area for climate management.

The contextualization of the relationship between artificial intelligence and climate hazard management is based on finding a common platform for a coherent positive relationship. Such a relationship should be based on facts and evidence about the importance of scientific cognition. The joint development of artificial intelligence for the management of climate hazards should be supported at the level of public administration dedicated to a servile role towards local and global communities. Political chaos in the area of understanding climate change is definitely not conducive to this. A common and consistent interpretation of climate hazards and a common range of systemic actions to build appropriate climate adaptation strategies are required.

7.3 Creating new climate products in the circular economy using artificial intelligence

New climate products can be the result of innovative circular value chains. The circular economy opens up many opportunities for the creation of new climate products, which are usually the result of designed and implemented climate business models. The circular economy, as a waste-free economy, is already climate-positive in its original assumptions. The circular economy can be considered a system which is capable of creating organizational and technical solutions throughout the life cycle of production and consumption that can serve as primary raw materials or be reused in the production process (Galvão et al. 2018, pp. 79–85, p. 79). The circular economy can be linked to the principles of a low-carbon economy and a green economy. A low-carbon economy is generally understood as an economy that produces a minimum level of greenhouse gas emissions (GHG). It is also defined as an economy characterized by activities with low CO_2 emissions to the atmosphere (Levy 2020). The term "green economy" was first used in the report entitled "Blueprint for a green economy", commissioned by the British government in 1989 (Pearce et al. 1989). The green economy shall be governed, among others, by the following principles:

- equality and integrity, both within and between generations;
- consistency with the principles of sustainability;
- a cautious approach to social and environmental impacts;
- recognition of natural and social capital (for example, by internalizing external costs), green accounting, the cost calculations of the product life-cycle and better management;
- sustainable and efficient use of resources, consumption and production;
- the need to adapt to current macroeconomic objectives by creating green jobs;
- poverty eradication, increased competitiveness and growth in core sectors (Europe's environment 2016, p. 97).

Table 17 below shows selected AI applications used to manage positive climate change.

It is necessary to pay attention to the intellectual strength of the specific solutions presented in the table, which are implemented on the basis of artificial intelligence. All this can have a synergistic and symbiotic effect, which gives a huge boost to effective and efficient climate hazard management.

Table 17. Selected AI applications used to manage positive climate change.

No.	Selected AI applications used to manage positive climate change	Description
1.	Endangered species conservation	Artificial intelligence helps to monitor populations of endangered species. For example, AI-powered drones are used in national parks to identify poachers and count animals, making it easier to control their numbers and health status.
2.	Air quality forecasting	AI algorithms analyze data from pollution sensors, predicting air quality in a given region. This helps react faster to deterioration of conditions and implement appropriate preventive measures.
3.	Smart waste management	Waste management is one of the most difficult challenges of modern civilization. Artificial intelligence can help to recycle, reduce waste and reduce its impact on the environment.
4.	Recycling process automation	AI-based vision systems are used in waste sorting plants to automatically recognize and segregate materials such as plastic, glass or metal. This makes recycling more efficient and less costly.
5.	Waste collection optimization	AI algorithms can analyze data on the filling of waste containers, optimizing the routes for their emptying. This reduces the fuel consumption of garbage collection vehicles and reduces exhaust emissions.
6.	Education and environmental awareness	Environmental education is crucial to building a sustainable future. Artificial intelligence supports this field by offering interactive tools and personalized learning experiences.
7.	Mobile applications supported by AI.	Smart learning applications teach users about green lifestyles, suggesting, for example, ways to save energy or reduce waste. By analyzing user behavior, applications can customize content and recommendations.
8.	Virtual environmental assistants	AI-based virtual assistants can answer questions related to ecology by educating users about environmental protection and promoting sustainable behavior in everyday life.

Table 17 contd. ...

...Table 17 contd.

No.	Selected AI applications used to manage positive climate change	Description
9.	Innovative green projects supported by AI	Artificial intelligence enables innovative green projects on an unprecedented scale: Rainforest Connection: An AI system that uses the sounds of rainforests to detect illegal tree felling. Microphones installed on trees analyze the sounds of chainsaws or heavy equipment and then notify the authorities. Google Earth Engine: A platform that uses AI to analyze huge sets of satellite data, which helps track environmental changes such as deforestation or soil degradation.
10.	Smart wind farms	Machine learning systems are used to optimize the performance of wind turbines by predicting wind speeds and adjusting the operation of equipment in real time.

Source: Own study based on https://soltech360.pl/2024/10/03/artificial-intelligence-in-service-ecology/

7.4 Analog operational security and physical protection systems, operational cybersecurity and cyber protection systems and climate change

Analog operational and physical security systems play a particular role in effective and efficient climate change management. Cybersecurity is primarily related to the mechanisms adopted for managing operational processes based on analog solutions, and physical protection is focused on the systemic security of operating systems and critical infrastructure against criminal physical attacks. Especially in case of significant threats, including military ones, the physical security of key facilities and critical infrastructure should be a priority strategic action. Then, it is crucial to define and designate control mechanisms for critical facilities and climate-critical infrastructure and to subject them to a specific supervisory process. Hybrid surveillance of physical and digital elements of systems and digital technological components should be systemic with high observation and measurement complexity, at the same time aimed at eliminating and limiting various types of interference occurring therein.

Operational cybersecurity and cyber protection systems play a particular role in effective and efficient climate change management.

Cybersecurity is primarily related to the mechanisms adopted for managing operational processes based on digital solutions, and cyber protection is focused on systemic security of ICT systems and critical infrastructure against attacks by hackers and cyberterrorists. In today's digital world, the cybersecurity ecosystem is a core pillar and foundation of protection against increasingly sophisticated threats. Automation, and in particular dedicated purposeful solutions such as SOAR (Security Orchestration, Automation and Response), definitely revolutionize the principles according to which individual organizations approach the application of cybersecurity logic and methodology (file:///C:/Users/Adam/Downloads/Security-Orchestration-Automation-and-Response-SOAR%20(1).pdf).

It should be remembered that industrial enterprises can often be treated as so-called high-climate-risk enterprises. The structure and core ontological entities of such organizations, i.e., business models, strategies, processes and projects, can be strongly supported by AI solutions.

7.5 Secure management of core climate data and artificial intelligence

Climate data are, among others, aggregated for the purposes of simulation calculation methods for predicting climate change. Climate data can be stored safely, for example, in so-called climate knowledge bases. A climate knowledge base is a place where valuable and up-to-date reports on climate change are collected and catalogued, enabling quick verification of the truth and scientific basis of information on climate change and global warming. Climate databases may include the following:

- Station databases,
- Grid databases,
- Synoptic map archives,
- Archive of synoptic cables,
- Archives of climatograms and aerological surveys,
- Satellite image archives,
- Climate simulations.

For meteorological weather purposes, these data may include the following:

- cloudiness, not defined as a decimal fraction from 0 to 1,

- dry thermometer temperature, expressed in °C,
- relative humidity, not defined as a decimal,
- direct solar irradiance to the horizontal surface, expressed in W/m2,
- diffuse solar irradiance, expressed in W/m2,
- wind speed, expressed in m/s,
- wind direction, expressed in degrees,
- total pressure, expressed in hPa,
- precipitation, expressed in mm/h.

It is important to know that the origin of climate data stems from different sources of data, information and knowledge. There are many tools for collecting historical data to manage and track climate data. The key here is to build interdisciplinary databases based on digital platforms, where data includes secondary analysis and data synthesis and detailed interpretation. It is important to convert weather content into analytical data. Longitudinal studies are also used here. Research into the dynamics of climate-related organizational and management processes should be longitudinal, as such studies search for mechanisms of climate change. It is very difficult to draw conclusions about the climate by looking at snapshots from the study of the most diverse organizations. It is more cognitively valuable to look at the functioning of the climate and how its parts are related from the perspective of their initial and final states. Longitudinal climate data research consists of those techniques, methodologies and activities that enable the observation, description and/or classification of organizational phenomena in such a way that processes can be identified and empirically documented. Longitudinal climate studies typically study processes over multiple periods. Since the duration between climate data collection activities is determined by the researcher, the number of climate data collection periods and their origin vary according to different patterns. Longitudinal studies are designed by changing six parameters, namely the length of the survey, the time between data collection activities, the number of data collection periods, the method of data collection, the research objectives and the subject matter of the analysis.

Core climate databases include those listed below in Table 18:

The databases presented above, of course, do not exhaust the potential of all dedicated climate databases, but they illustrate their significance in the processes of effective climate management.

Table 18. List of core climate databases.

No.	Core Climate Databases
1.	Berkeley Earth Data https://berkeleyearth.org/data/#:~:text=Berkeley%20Earth%20provides%20high-resolution%20land%20and%20ocean%20time
2.	Climatic Research Unit https://www.uea.ac.uk/groups-and-centres/climatic-research-unit
3.	Global Historical Climatology Network https://www.ncei.noaa.gov/products/land-based-station/global-historical-climatology-networkmonthly#:~:text=The%20Global%20Historical%20Climatology%20Network%20monthly%20(GHCNm)%20dataset
4.	A monthly global paleo-reanalysis (input) https://www.nature.com/articles/sdata201776#:~:text=This%20data%20set%20combines%20the%20advantage%20of%20traditional#:~:text=This%20data%20set%20combines%20the%20advantage%20of%20traditional
5.	The International Surface Pressure Databank https://rmets.onlinelibrary.wiley.com/doi/10.1002/gdj3.25#:~:text=The%20International%20Surface%20Pressure%20Databank%20%28ISPD%29%20is%20the,collection%20of%20global%20surface%20and%20sea-level%20pressure%20observations
6.	International Surface Temperature Initiative https://rmets.onlinelibrary.wiley.com/doi/10.1002/gdj3.8#:~:text=This%20databank%20release%20forms%20the%20foundation%20for%20the
7.	The McGill DRAW (Data Rescue: Archives and Weather) project shows how climatologists, geographers, archivists, data scientists, and coders together built a citizen-science-based transcription platform to transform the McGill Observatory paper records into a traceable and sustainable database.
8.	Climate Data Rescue, https://github.com/opendata-rescue/climate-data-rescue Data Rescue: Archives and Weather, https://citsci.geog.mcgill.ca/en/
9.	Historical climate observations in Canada: 18th and 19th century https://rmets.onlinelibrary.wiley.com/doi/full/10.1002/gdj3.11#:~:text=Daily%20observations%20of%20weather%20and%20climate%20for%20the
10.	ACRE South Africa https://www.myacreofafrica.org/#:~:text=My%20Acre%20of%20Africa%20has%20the%20potential%20to
11.	Early instrumental meteorological observations in Switzerland https://essd.copernicus.org/articles/12/1179/2020/#:~:text=We%20describe%20a%20dataset%20of%20recently%20digitised%20meteorological
12.	Deutscher Wetterdienst https://www.surveyxact.de/servlet/com.pls.morpheus.web.pages.CoreRespondentCollectLink Anonymous
13.	European Climate Assessment & Dataset https://www.ecad.eu/#:~:text=Welcome%20to%20the%20website%20of%20the%20European%20Climate

Table 18 contd. ...

...Table 18 contd.

No.	Core Climate Databases
14.	Early meteorological records from Latin America and the Caribbean during the 18th and 19th centuries https://europepmc.org/article/PMC/5685157
15.	Improved Understanding of Past Climatic Variability from Early Daily European Instrumental Sources https://cordis.europa.eu/project/id/ENV4970511#:~:text=Objectives%20:%20To%20assess%20correction%20and%20homogenisation%20protocols
16.	Historical Instrumental Climatological Surface Time Series of the Greater Alpine Region https://rmets.onlinelibrary.wiley.com/doi/abs/10.1002/joc.1377#:~:text=This%20paper%20describes%20the%20HISTALP%20database,%20consisting%20of
17.	Japan-Asia Climate Data Program https://ap-plat.nies.go.jp/#:~:text=AP-PLAT%20is%20an%20online%20information%20platform%20that%20bridges
18.	C3-EURO4M-MEDARE Mediterranean historical climate data https://rmets.onlinelibrary.wiley.com/doi/full/10.1002/gdj3.12#:~:text=Title:%20C3-EURO4MMEDARE%20Mediterranean%20historical%20climate%20data.%20Authors:%20Centre
19.	HARD 2.0 - Historical Arctic Database http://www.hardv2.prac.umk.pl/#:~:text=This%20website%20provides%20Historical%20Climate%20Data%20from%20the
20.	SEA-historical climate data https://www.ncei.noaa.gov/cdoweb/#:~:text=Climate%20Data%20Online%20(CDO)%20provides%20free%20access%20to
21.	Southeast Asian Climate Assessment & Dataset http://sacad.bmkg.go.id/#:~:text=SAOBS%20is%20a%20daily%20gridded%20observational%20dataset%20for
22.	Two-Century Precipitation Dataset for the Continent of Africa
23.	Two-Century Precipitation Dataset for the Continent of Africa https://journals.ametsoc.org/view/journals/bams/93/8/bams-d-11-00212.1.xml#:~:text=Documentary%20information,%20hydrologic%20indicators,%20and%20rain%20gauge%20records
24.	Forts and Volunteer Observer Database https://mrcc.purdue.edu/FORTS#:~:text=This%20web%20site%20provides%20information%20about%20these%2019th

Source: Own study.

7.6 The Climate-Changing Digital Twin as a new solution to climate hazards

Global climate models provide simulations of climate change on a scale that is too large to be used to assess acute hazards to specific buildings and even cities – these models generate information in a grid with coordinates about sixty miles away. Researchers are switching to higher-resolution GCM models, but a significant limitation of this approach is the computing power and time required to run the model, as the data becomes more detailed and therefore more cumbersome (https://www.gfdl.noaa.gov/high-resolution-atmosphere-modeling/).

The concept of the digital twin of the Earth assumes the convergence of Big Earth Data with physics-based models in an interactive computing environment that allows for monitoring and predicting environmental and social disorders for use in sustainable management (Li et al. 2023). The Climate-Changing Digital Twin (Climate DT) is a new type of climate information system funded by The Destination Earth program, which is focused on assessing the impact of climate change and various adaptation strategies at local and regional levels from a global perspective using a strategy that is based on user demand addressed to the production chain. Its solution is an extension and application model resulting from the classic digital twin. The origins of the digital twin are attributed to Michael Grieves and his collaboration with NASA's John Vickers, with Grieves presenting the concept during a lecture on product lifecycle management in 2003 (Grieves 2014). The description of the concept is based on the digital twin as a complete virtual description of the physical product, which is accurate at both micro and macro levels. A Digital Twin Prototype refers to a virtual description of a product prototype, containing all the information required to create a physical twin. A Digital Twin Instance is associated with a specific instance of a physical product that remains associated with an individual product throughout the life of that product.

A Digital Twin Aggregate is a combination of all instances of the digital twin. The digital twin environment is a multi-domain space for physics applications to support digital twins. These operations include predicting performance and searching information (Jones et al. 2020). The digital twin must cover the entire life cycle of its physical goals. It is incorrect to say that the digital twin exists only when there is a physical artifact. There is tremendous value in using a digital twin before a physical product is created. The ideal is virtual product creation, virtual product testing, virtual product manufacturing and virtual product support. Only when we do everything right do we move physical atoms to create a physical product (Grieves 2023). Based on the overall concept of the digital twin, a Climate-Changing Digital Twin incorporates a multi model of global climate simulations with no-percentage resolution and relies on an immediately quantified qualitative assessment. What is important is that it takes into account the importance of climate and non-climate media. It integrates large amounts of relevant climate and non-climate data in the area of research and development. The logic of this solution is based on streaming containing the output model enabled in the standard formula (general generic vector state, GSV interface: new access to the Climate

Digital Twin data). The GSV interface includes standardized output, a list of variables, horizontal grids, time frequencies, and pressure levels. Access to data for internal and external users is based on the so-called data bridge. With the Climate-Changing Digital Twin (Climate DT), data customers have access to the full state of the model, which is based on interactivity and scalability. Interactivity focuses on the development of enabled simulations containing on-demand variables. Scalability, on the other hand, includes adding new applications and specific requirements for them. Scalability means ensuring increasingly more efficient work as the number of components increases. This is, among others, a feature of computer networks consisting in the ability to continuously expand them. Scalability is sometimes defined as "the ease with which a system or component can be modified depending on the type of problem". A scalable system has three simple features, namely:

- The system can adapt to increased use.
- The system can accommodate larger amounts of data.
- The system is easy to use and works with reasonable efficiency.

Scalability is not just about speed. System efficiency and scalability vary and mutually correlate. Efficiency measures how quickly and efficiently a system can perform certain accounting tasks, while scalability measures the trend of efficiency under increased load (https://www.comp.nus.edu.sg/~seer/book/2e/Ch06.%20Scalability.pdf).

Daniel A. Menascé and Virgilio A.F. Almeida argue that a system is scalable if there is a "simple" way to update the system to support increased trade while maintaining proper efficiency. Simple means that no changes to the system architecture or software should be required to scale the system (Menascé and Virgilio Almeida 2000, p. 10–15). The Universal Scalability Law (USL) in computer science is a model for predicting the scalability of hardware and software. It uses system performance as a load function to predict scalability for the system. The USL function is used to create a model from a formula and data frame. As a result, the USL model produces two coefficients: sigma models the theorem, and kappa delays the coherence of the system. The Universal Scalability Law was formulated by Neil J. Gunther (Gunther 2007, Gunther 2008). The following are core theorems and mathematical formulae describing this law. The system performance is expressed as deviation from linear scalability. The Universal Scalability Law is used to predict the scalability of both hardware and software. The USL model works with one independent variable (e.g., virtual users, processes, threads) and one dependent variable (e.g., bandwidth). Therefore, the formula of the model must be in a simple "response-predictor" format. The coef function extracts the coefficients from the model object. The argumentative method selects which solution is to be applied to the model solution, namely:

- "default" for the default value method using a two-stage polynomial transformation. It can only be used if the model frame contains value for normalization, where the converter is equal to "1" for one measurement. This is an algorithm introduced by Dr. Neil J. Gunther in his book entitled *Guerrilla Capacity Planning*;
- "nls" for a non-linear regression model. This method evaluates not only the sigma and kappa coefficients, but also the scale. The nls normalization factor with the port algorithm is used internally to solve the model. Therefore, all limitations of the "port" algorithm apply;
- 'nlxb' with a non-linear regression model uses the nlxb function from the nlmrt cable. This method also evaluates both the coefficients and the normalization factor. It is expected to be more robust than the nls method.
- The USL (Universal Scalability Law) combines the following important effects:

in one scalability model that determines the relative capacity of C(N):

$$C(N) = \frac{N}{1 + \alpha\,(N-1) + \beta\,N\,(N-1)} \tag{1}$$

The three concepts in the denominator of equation (1) are related to the three Cs, respectively:

1. A concurrency level (...) or ideal parallelism: Basically, linear scaling.
2. A contention level (of α strength) due to waiting for common resources.
3. A coherency level due to delay in data harmonization (...) due to point-to-point exchange.

These parametric values are defined in the range of: $0 \leq \alpha, \beta < 1$. The independent variable N can represent software scalability, where the number of users or load generators (N) increases according to the fixed hardware configuration. In this case, the number of users acts as an independent variable, while the CPU configuration remains constant for a given range of user load meters. This is the most common situation in the load testing environment, where tools such as LoadRunner HP or Apache JMeter are used.

In the case of hardware scalability, the number of physical processors (N) increases in hardware configuration at the same time, maintaining a constant user load on the CPU. In this case, it is assumed that the number of users per CPU (e.g., 100 users per CPU) remains the same for each added processor. For example, a 32-processor platform would use a load of N = 3200 users for the test platform.

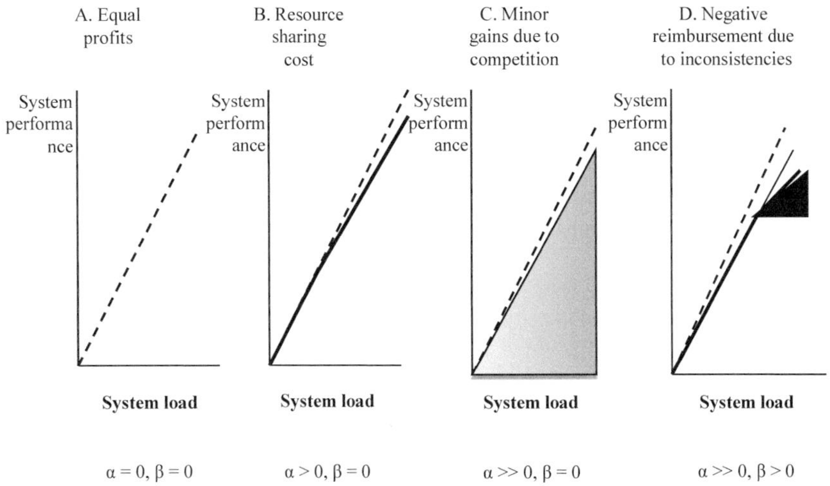

Figure 20. Core USL effects (universal scalability law). Source: http://www.perfdynamics.com/Manifesto/USLscalability.html.

Equation (1) says that the scalability of hardware and applications are two sides of the same coin: something that is not generally recognized. The non-zero value β is related to the measured capacity, which regresses, i.e., decreases with increasing load or platform configuration.

Note: The purpose of using equation (1) is NOT to produce a curve, which passes through each data point. It is called curve matching, which is what graphic artists do with protrusions. As von Neumann said: '**With four parameters I can fit an elephant, and with five I can make him wiggle his trunk!**" (I only have two at least).

Matched coefficients in equation (1) provide an assessment of the user load where maximum scalability occurs for D:

$$N_{max} = \sqrt{1-\alpha/\beta} \tag{2}$$

and the corresponding capacity X_{max} at load N_{max} is given by:

$$X_{max} = X(1) \times C(N_{max}) \tag{3}$$

The scalability profiles B and C (above) correspond to β = 0, which means that N_{max} occurs in infinity. In other words, the scalability curve simply approaches the ceiling at 1/α (horizontal dashed line on the charts).

The scalability of IT applications will be understood as the ability to adjust the application to the size of tasks in the organization (the size of the organization). Therefore, it is a size that characterizes the ability of the application to perform tasks with a different range of intensity of these tasks; in simple terms, the maximum and minimum number of elements at which the application is useful (working with acceptable performance) (Mazur 1999). The Climate Digital Twin concept includes three main parts, namely:

1. physical products in the real space,
2. virtual products in the virtual space and
3. data and information connections that connect virtual and real products.

7.7 Conclusions

The contextualization of the place and role of artificial intelligence in climate hazard management is fundamental to climate outcomes in the short and long term to address climate hazards. Only the evidence-based and comprehensive use of the technological potential of artificial intelligence becomes a determinant for improving climate security at the level of natural, social, technological and economic ecosystems. Such logic and its implementation in a systemic way should ensure that decision-makers' decisions are directed towards creating solutions conducive to positive climate change.

References

Europe's environment. An Assessment of Assessments. (2016). 3 Green economy, p. 97.

file:///C:/Users/Adam/Downloads/Security-Orchestration-Automation-and-Response-SOAR%20(1).pdf.

Galvão, G. D. A., de, Nadae J., Clemente, D. H., Chinen, G., and Monteiro de Carvalho, M. (2018). *Circular Economy Overview of Barriers*. Procedia CIRP, 73: 79–85.

Grieves, M. W. (2023). *Digital Twins: Past, Present, and Future in: The Digital Twin*, Noel Crespi. Adam T. Drobot, Roberto Minerva, Editors, Springer Nature Switzerland AG, https://doi.org/10.1007/978-3-031-21343-4.

Grieves, M. (2014). *Digital twin: manufacturing excellence through virtual factory replication.*

Gunther, N. J. (2007). *Guerrilla Capacity Planning: A Tactical Approach to Planning for Highly Scalable Applications and Services*. Springer, Heidelberg, Germany, 1st edition.

Gunther, N. J. (2008). *A general theory of computational scalability based on rational functions.* CoRR, abs/0808.1431.

http://www.perfdynamics.com/Manifesto/USLscalability.html.

https://www.eea.europa.eu/publications/europes-environment-aoa/chapter3.xhtml.

https://www.gfdl.noaa.gov/high-resolution-atmosphere-modeling/

https://www.comp.nus.edu.sg/~seer/book/2e/Ch06.%20Scalability.pdf

Jones, D., Snider, C., Nassehi, A., Yon, J., and Hicks, B. (2020). *Characterising the Digital Twin: A systematic literature review*, CIRP Journal of Manufacturing Science and Technology 29, p. 36–52, https://doi.org/10.1016/j.cirpj.2020.02.002.

Levy, C. (2020). *A Low Carbon Economy – A Knowledge Economy Programme Report*. The Work Foundation, April 2010.

Li, X., Feng, M., Ran, Y., Su, Y., Liu, F., and Huang, C. (2023). *Big Data in Earth system science and progress towards a digital twin*. Nat. Rev. Earth Environ., 4: 319–332 (2023). https://doi.org/10.1038/s43017-023-00409-w.

Mazur, D. (1999). *Skalowalność w systemach informatycznych zarządzania* [Scalability in management information systems], Zeszyty Naukowe Politechniki Śląskiej.

Menascé, D. A., and Virgilio Almeida, A. F. (2000). *Challenges in scaling e-business sites*, Proc. 2000 Computer Measurement Group Conference, Orlando, FL, s.10–15.

Pearce, D., Markandya, A., and Barbier, E. B. (1989). *Blueprint for a Green Economy*. Earthscan, Londyn.

Chapter 8

The Authors' Original Concept for the Development of the Evolutionary Cycle of Climate Business Models

From Sustainable Business Models to AI Climate-Changing Business Models

8.1 Introduction

The aim of this chapter is to present the authors' original concept for the development of the evolutionary cycle of climate business models – from sustainable business models to AI Climate-Changing Business Models.

The scope of the chapter covers issues discussed from sustainable business models to AI Climate-Changing Business Models.

The subject of the chapter deals with the evolutionary cycle of climate business models.

Based on literature studies, including a systematic literature review as well as a critical literature review, variables affecting the creation of the authors' concept of development from the evolutionary cycle of climate business models, which was juxtaposed from an orthodox approach based on sustainable business to a positivist approach associated with the creation of climate-changing business models, were analyzed. The variability of parameters and the complexity of the economic system are conducive to the creation of models that are gradually evolving as technological progress increases. Such progress cannot be stopped and solutions must be adapted to modern needs. The diagram presents the processes of digital transformation, energy transition and climate transition based on the assumptions of artificial intelligence mechanisms

DOI: 10.1201/9781003434016-8

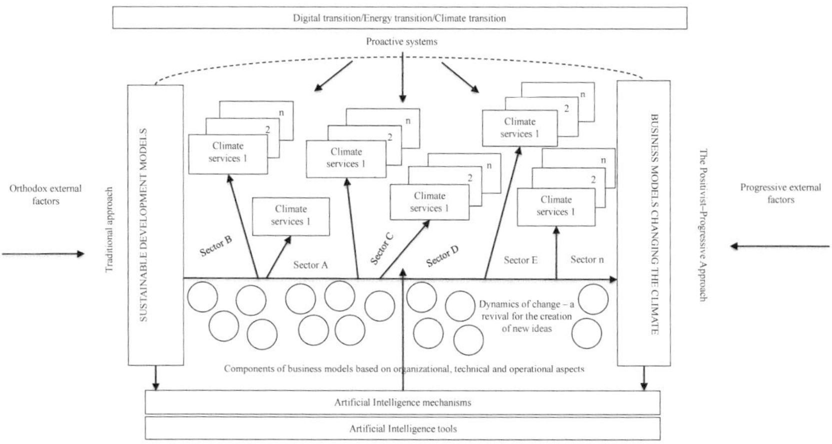

Figure 21. A proactive business development model based on progressive Climate-Changing Business Models. Source: Own study.

and appropriate tools for this purpose. In this way, a shift occurred from sustainable development models to Climate-Changing Business Models. Attention was also paid to the role that climate services, which are relevant to specific sectors of the economy and some of which are interdisciplinary, play. The components of the business model therefore move from sustainable development models to Climate-Changing Business Models and are stimulated by the mechanism of artificial intelligence and the functionality of climate services. The management model configured in this way fulfills the contemporary requirements of monitoring and measuring factors affecting climate change and, consequently, the quality of life of citizens and the development of the economy. The corresponding figure is presented below.

To create such a model, it is necessary to understand the complexity of such systems that are important for creating business ecosystems in which other business models are created. Creating Climate-Changing Business Models is not easy. It is not only a question of ensuring that the environmental impact is neutralized or minimized at worst, but of finding innovative solutions that will break the rules and principles of the economic system according to Schumpeter's creative destruction concept and allow the creation of such conditions that anomalies, dysfunctions and other worrying phenomena related to climate change will be halted, and the synergy effect between all groups of companies operating in economic sectors around the world will create the so-called green scalability, i.e., a situation in which economic activity will repair them instead of adversely affecting the environment.

8.2 Conclusions

Building a proactive business development model based on progressive Climate-Changing Business Models seems to be an interesting analytical–conceptual approach to linking the principles of digital transition, energy transition and climate transition for effective and efficient climate hazard management by means of artificial intelligence. The logic and methodology built in this model approach can give great substantive value to mitigate and neutralize climate hazards along with the creation of positive climate changes.

Chapter 9

The Authors' Original Concept of Climate Hazard Management by Means of Artificial Intelligence

9.1 Introduction

The aim of this chapter is to present the authors' original concept of climate hazard management by means of artificial intelligence.

The scope of the chapter covers issues related to the artificial intelligence model in climate hazard management and is based on four pillars, namely natural climate hazards (NZK1, NZK2,......NZKn), human-induced climate hazards (CZK1, CZK2,..... CZKn), climate anomalies (AK1, AK2,.....AKn) and climate dysfunctions (DK1, DK2, ...DKn).

The subject of the chapter is to present the authors' concept of climate hazard management using artificial intelligence in relation to climate data processed by artificial intelligence.

The artificial intelligence model in climate hazard management presented above in Figure 21 is based on four pillars, namely:

1. Natural climate hazards (NZK1, NZK2,......NZKn).

   Natural climate hazards can be interpreted as those climate hazards which man has no influence on and can only skillfully neutralize and mitigate (Table 19).

2. Human-induced climate hazards (CZK1, CZK2,..... CZKn).

   Human-induced climate hazards can be interpreted as such hazards that are directly or indirectly caused by human existential, social and economic activities (Table 20).

DOI: 10.1201/9781003434016-9

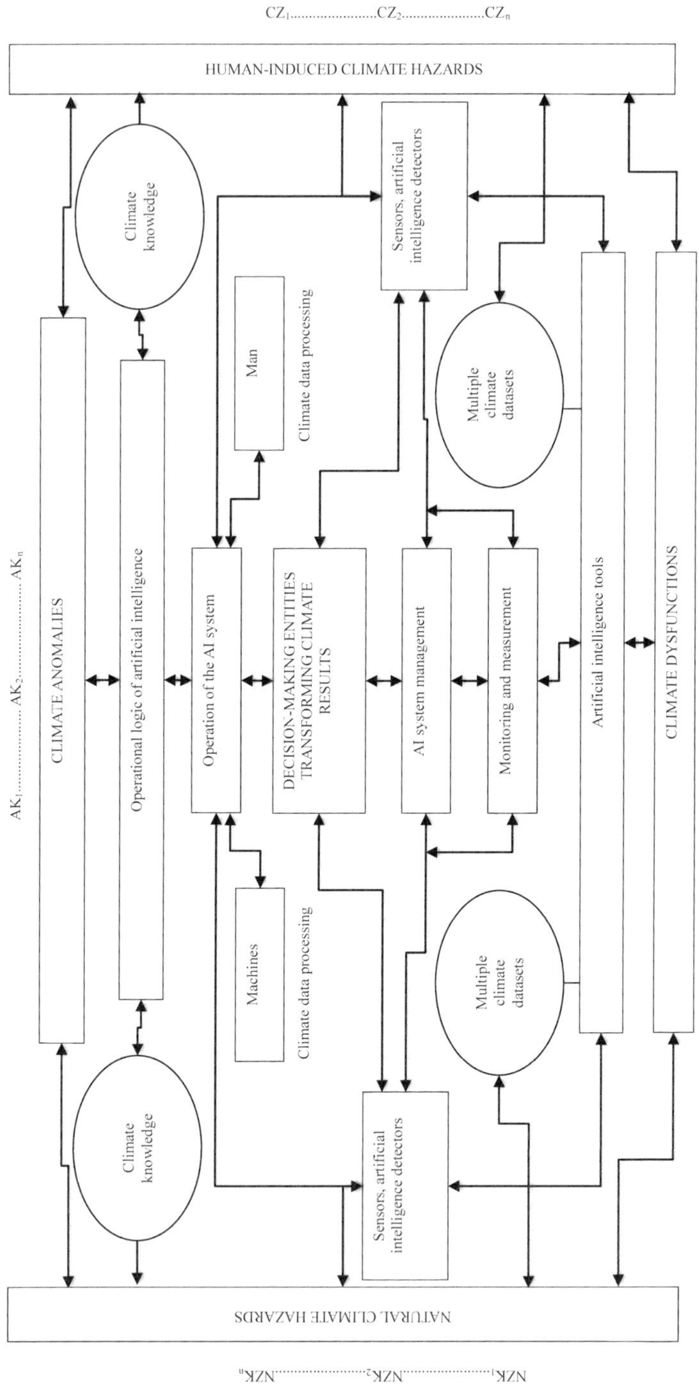

Figure 22. Artificial intelligence model in climate hazard management. Source: Own study.

Table 19. Natural climate hazards (NZK1, NZK2,......NZKn).

No.	1. Natural climate hazards (NZK1, NZK2,......NZKn).
1.	Warming
2.	Heat waves
3.	Precipitation
4.	Droughts
5.	Floods
6.	Fires
7.	Storms/hurricanes
8.	Sea level rise
9.	Changes in vegetation
10.	Changes in ocean water chemistry

Source: Own study.

Table 20. Human-induced climate hazards (CZK1, CZK2,..... CZKn).

No,	Human-induced climate hazards (CZK1, CZK2,..... CZKn).
1.	Combustion of fossil fuels in the energy sector resulting in carbon dioxide (CO2) emissions
2.	Carbon-emitting transport (CO_2)
3.	Carbon-emitting industry (CO_2)
4.	Deforestation in forests that absorb CO2, reducing the Earth's ability to regulate greenhouse gas levels
5.	Atmospheric pollution: Industrial emissions and the incineration of garbage release toxic substances and particles into the atmosphere
6.	Production and consumption
7.	Excessive use of natural resources increases CO_2 emissions and generates waste that affects the environment

Source: Own study.

3. Climate anomalies (AK1, AK2,.....AKn).

   Climate anomalies can be interpreted as such anomalies that are different from the changes occurring in the type of anomaly itself and/or in places where they have never occurred. Usually these are sudden and/or short-term phenomena (Table 21).

4. Climate dysfunctions (DK1, DK2, ...DKn).

   Climatic dysfunctions are such climatic factors that usually occur as an accumulation of surprising and abnormal climatic phenomena that usually have a long-term effect (Table 22).

Table 21. Climate anomalies (AK1, AK2,.....AKn).

No.	Climate anomalies (AK1, AK2,.....AKn).
1.	Heat and heatwaves
2.	Excessive rainfall
3.	Long periods of no rainfall
4.	Extreme atmospheric phenomena such as tornadoes
5.	Unusual winter conditions, such as snowfall in places where it usually does not snow, or vice versa – no snow in places where it is expected
6.	Sudden changes in atmospheric pressure
7.	Weather anomalies can have a significant impact on people's lives, the economy and the environment

Source: Own study.

Table 22. Climate dysfunctions (DK1, DK2, ...DKn).

No.	Climate dysfunctions (DK1, DK2, ...DKn).
1.	Melting of glaciers, which will raise sea and ocean levels
2.	Increased meteorological extremes, especially the strength and frequency of hurricanes and episodes of drought and heat
3.	Changes in the intensity and predictability of precipitation, which may cause local flooding in addition to drought
4.	Threat to the existence of species particularly vulnerable to climate change
5.	Melting of snow cover, which is the main source of drinking water for social groups
6.	Difficulties in cultivation and breeding leading to reduced food supply
7.	Ocean acidification leading to a change in biological balance and dominance of unfavorable species (e.g., algae) at the expense of other animals and plants
8.	Soil degradation leading to desertification, avalanches and loss of biodiversity

Source: Own study.

The climate data processed by artificial intelligence in these four pillars is used to make core strategic and tactical–operational decisions by decision-makers in the area of climate and climate change management. AI solutions are implemented on the basis of the operational logic of artificial intelligence adopted in this topic. Historical climate data and climate knowledge are processed. They are obtained from secondary data and through appropriate artificial intelligence sensors and detectors. The operational logic of artificial intelligence is based on action in the machine–human relationship. Decision-making entities, on the other hand, manage the AI system and implement AI monitoring and measurement processes based on processed climate data resulting from multiple climate data sets. For this purpose, dedicated, specialized AI tools are used.

9.2 Conclusions

The AI model defined in this way in climate hazard management is a comprehensive solution, as it refers to different types of climate hazards with different mechanisms of impact. Only a full and integrated climate hazard management strategy can underpin systemic climate hazard management. Then, the resulting final effect guarantees the safe functioning of various types of ecosystems.

Chapter 10

The Authors' Original Concept of Developing AI Mechanisms and Solutions in Climate Hazard Management

10.1 Introduction

The aim of this chapter is to present the authors' original concept of developing AI mechanisms and solutions in climate hazard management.

The scope of the chapter covers issues related to the definition of two areas of application of artificial intelligence – the climate and the mechanisms of using artificial intelligence.

The subject of the chapter concerns the authors' original arrangement of concepts defining the boundaries that climate researchers should deal with in the context of business models.

Solving scientific problems with scientific methods should be based on the concept of evidence-based practice, and in particular on the assumptions of evidence-based management (Barends and Rousseau 2018, Jabłoński et al. 2025). Only evidence created in the course of analyses of the systematic literature review can capture these core concepts, words and sentences on the basis of which the most modern solutions in the field of theory of practice can be inferred and sought. The presented research based on a systematic literature review allowed for the identification of numerous key concepts that should form the basis for putting them into a uniform coherent system defining the boundaries that researchers should deal with in the sphere of the impact of artificial intelligence on the search for solutions for mitigating hazards and eliminating climate change. This task was undertaken by the authors of this book, who detailed the following set of concepts shaping the ability of researchers to identify cognitive spaces. A collection of these concepts is given in Figure 22.

DOI: 10.1201/9781003434016-10

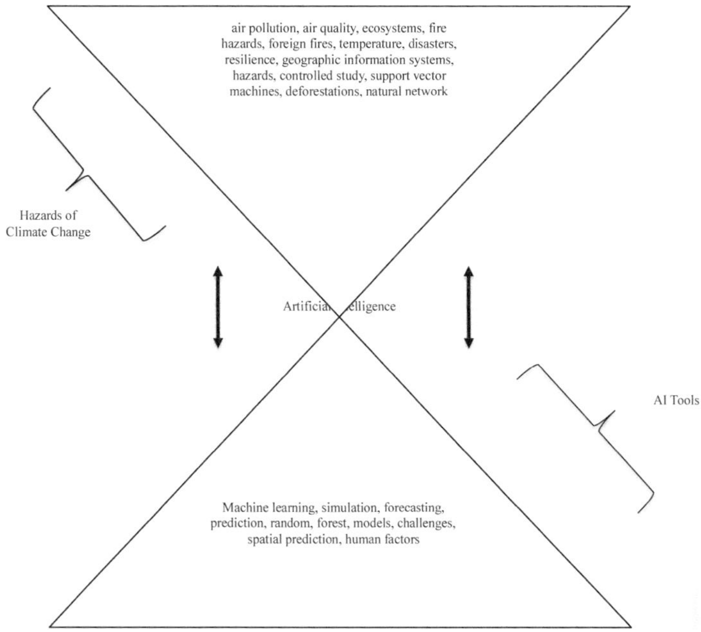

Figure 23. Two AI application spaces – a climate area and artificial intelligence usage mechanisms area. Source: Own study based on a systematic literature review described in chapter 2.

Thus, an original arrangement of concepts defining the boundaries that climate researchers should deal with in the context of business models was created.

10.2 Conclusions

The two AI application spaces presented – the climate area and the area of mechanisms of using artificial intelligence – combine the logic of climate hazard management with machine learning solutions, simulation, forecasting, prediction, randomness, climate models, climate challenges, spatial prediction and the human factor into one coherent system.

Such a multidimensional approach can have a significant number of results with a high climate impact.

References

Barends, E., and Rousseau, D.N. (2018). *Evidence – Based Management*, Kogan Page Limited, New York.

Jabłoński, M., Jabłoński, A., Janulek, P., Dulęba, D., and Glenszczyk, M. (2025). *Traktat o zasadach zarządzania dowodowego - przyszłość zarządzania* [Treaty on the Principles of Evidence-Based Management - The Future of Management], CeDeWu, Warszawa.

Chapter 11

The Authors' Original Spiral of Knowledge on the use of Artificial Intelligence Mechanisms to Solve Problems in the Field of Climate Hazards and Climate Change

11.1 Introduction

The aim of this chapter is to develop the authors' original knowledge spiral on the use of artificial intelligence mechanisms to solve problems in the field of climate hazards and climate change.

The scope of the chapter covers issues related to determining the knowledge spiral in the use of AI in monitoring and neutralizing adverse climate change, including mitigation of hazards related to weather and climate anomalies related to various sectors of the economy such as agriculture, tourism, etc.

The subject of the chapter concerns the use of a modified knowledge spiral, which seems justified in the case of solving AI-related problems.

The systematic review of the relevant literature in this monograph allowed for the specification of keywords related to the issues of artificial intelligence, its impact on the creation of climate hazards, as well as artificial intelligence monitoring climate change. Understanding concepts closely related to artificial intelligence and its impact on climate change allowed for the identification of such key concepts that can be used to create a universal model describing the current state of science and knowledge in the field of strategic information management systems and productive analyses used to capture climate anomalies and weather

DOI: 10.1201/9781003434016-11

changes, including all kinds of other variables that can affect the quality of life of people and the entire natural ecosystem. Research based on the evidence-based practice concept, i.e., research that can be used by decision-makers to make effective and efficient decisions in the climate sphere, is becoming a crucial challenge for researchers trying to find solutions that will halt the adverse impact of climate change on natural ecosystems, the quality of life of the world's inhabitants and anomalies threatening human lives. The resulting model of a semantic map creating dependencies, among others, created the opportunity to indicate those thematic areas for which decisions should be made in the first place, in order to create new conditions for human life and the creation of safe natural ecosystems. Such an understanding of science based on quantitative and qualitative research, including randomized research on specific population groups, should give decision-makers a chance to make better decisions, in particular those that will make use of all scientific achievements to date. The issue is so serious that it concerns not only our generation, but future generations as well, where people should live in better conditions and enjoy their potential that the nature gives them. In this regard, it is necessary to develop the knowledge and awareness of societies about the place and role of digitalization, including artificial intelligence, in relation to the creation of positive climate change. Figure 24 shows a spiral of knowledge

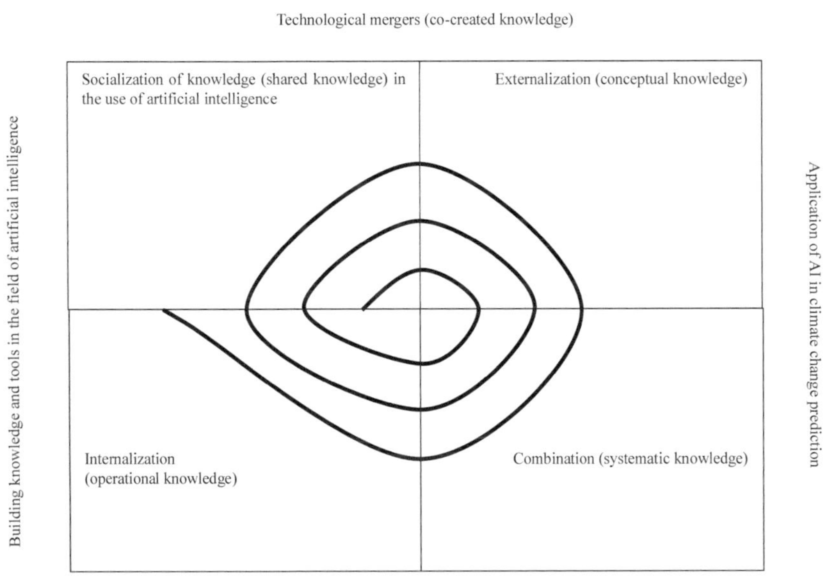

Figure 24. Spiral of knowledge on the use of AI in monitoring and neutralizing adverse climate change, including mitigation of weather and climate anomaly hazards related to various sectors of the economy such as agriculture, tourism, etc. Source: Own study based on Nonaka and Takeuchi (1995).

on the use of artificial intelligence mechanisms to solve problems in the field of climate hazards and climate change.

The use of a modified knowledge spiral seems appropriate for solving problems related to artificial intelligence. This is particularly important in difficult issues such as climate change. After all, knowledge is created by artificial intelligence, which is adopted for knowledge. It aims to speed up decision-making processes and help solve difficult and complex problems. This knowledge was created as a result of the integrative modeling of artificial intelligence to support scientists in the search for new solutions that generate progress.

11.2 Conclusions

Modelling the use of knowledge to shape a positive climate impact seems appropriate for generating multidimensional results in the area of climate change. The knowledge built into and used to achieve climate results should be created in a spatial approach. Only then will the increase in knowledge be able to be a dynamic accelerator of climate impact in various fields. This comprehensive approach is a positive determinant of climate change.

References

Nonaka, I., and Takeuchi, H. (1995). *The Knowledge – Creating Company, How Japanese Companies Create the Dynamics of Innovation*, Oxford University Press, New York, Oxford.

Chapter 12

Discussion of Research Results

Conclusions for Theory, Practice, Future Research Directions and Limitations of the Research Process, Theoretical Implications, Practical Implications, Potentially Interesting Future Research Directions, Research Limitations

12.1 Introduction

The purpose of this chapter is to present the results of the discussion on research results – conclusions for theory, practice, future research directions and limitations of the research process, theoretical implications, practical implications, potentially interesting future research directions and research limitations.

The scope of the chapter covers issues related to the identification of core conclusions for theory, practice, future research directions and limitations of the research process, constituting a conclusion of the monograph.

The subject of the chapter concerns the identification of the most important theoretical aspects and practical implications, and the identification of potentially interesting future research directions and research imitations.

The issue of artificial intelligence has a very wide application and is analyzed from numerous points of view. So far, the relevant literature has not included the discussion, analyses, concepts, ideas and mechanisms of applying artificial intelligence in the context of climate change conditions. This scientific monograph satisfies the expectations of recipients interested

DOI: 10.1201/9781003434016-12

in how artificial intelligence can support decision-makers and also citizens in the processes of neutralizing climate change, which increasingly leads to adverse events such as weather anomalies, unusual natural phenomena, environmental pollution, restless behavior of the Earth, including melting glaciers, volcanic eruptions, etc. The research and analysis carried out in this monograph allowed for the formulation of a number of conclusions and covered a significant number of areas that are the starting point for further research and exploration of this topic by other scientists. To this end, several leading thematic trends have been defined, describing both theoretical and practical issues examined in individual chapters of the book.

The theoretical part of the work included:

- Artificial intelligence as a research object in management sciences and climate sciences and as an application tool for modern business solutions.
- A systematic literature review – a description of the landscape of AI research.
- Key meanings and interpretations of artificial intelligence in the context of climate change.
- Artificial intelligence and an overview of selected solutions for creating positive changes for the climate.
- Climate hazard management mechanism by means of artificial intelligence solutions.
- Contextualization of the place and role of artificial intelligence in climate hazard management.

The practical part of the book included:

- The authors' original concept for the development of the evolutionary cycle of climate business models – from sustainable business models to AI Climate-Changing Business Models.
- The authors' original concept of developing AI mechanisms and solutions in climate hazard management.
- The authors' original concept of climate hazard management by means of artificial intelligence.
- An original spiral of knowledge of the use of AI mechanisms to solve problems in the field of climate hazards and climate change.

In the course of research and analysis, several core theoretical conclusions can be drawn:

1. Knowledge of the applications of artificial intelligence is pre-paradigmatic; so far no uniform and agreed approach, which has also been approved by the scientific community, has been developed,

which could be the beginning of a full theory of cognition that can be considered a paradigm. There is still a great deal of space to learn and conduct a lot of scientific research in both technical and social spheres.

2. In technical terms, the concept of artificial intelligence developed on the basis of the assumptions of deep machine learning solutions. The following trends emerged: the development of machine learning, the world of neural networks, the Hopfled network and the Bolzmann machine, convulsive learning, neural information processing system, the age of algorithms or deep intelligence as a systemic approach to these issues. It should be mentioned that the usefulness in the context of the assumptions for the discipline of management and quality sciences, especially additionally in terms of the determinants of Climate-Changing Business Models, is not yet sufficiently mastered so that is could be simply described and solutions unequivocally accepted in the environment could be determined. These theories are yet to be developed against the background of various concepts and research pertaining to a wide range of management issues and dedicated to climate change monitoring.
3. There is a need to search for answers to the questions of whether enterprises powered by artificial intelligence can exist and how their business models will be shaped, but also how the organizational structures of such organizations will look. Artificial intelligence thus changes the principles of understanding current canons in the sphere of work organization, the course of business and post-operational processes, the principles of project management, the role of the human factor against the background of the value provided by artificial intelligence solutions, but also the creation of trust and relationships with customers towards AI solutions.
4. In the context of artificial intelligence in the face of a global phenomenon such as climate change, in many cases theories must meet unexpected weather phenomena and anomalies not recognized by the world of science, which further complicates the assumptions, principles and ways of building commonly accepted theories in this area.

In the course of research and analysis, several core practical conclusions can be drawn:

1. Every day new and better tools that use artificial intelligence engines are being created.
2. Progress in the development of the application and widespread use of artificial intelligence is characterized as exponential.
3. After the implementation of artificial intelligence in business and social life, the world is no longer the same.

4. It is necessary to search for positive AI applications that will improve the quality of human life and will be the starting point and direction for creating new solutions, especially in terms of saving the natural environment, anticipating adverse events for the life of natural ecosystems and the location of current human life and understanding of progress to inhibit some activities, the impact of which on the natural environment is adverse and destroys what has been developed by man through technological progress and, due to excessive consumption, may be shifted to subsequent generations.

Potentially interesting directions for future research should certainly include the association of artificial intelligence with the concepts of evidence-based practice, evidence-based policy and in particular evidence-based management (Barends and Rousseau 2018, Jabłoński 2025). Linking codified knowledge to databases used in AI algorithms can significantly affect the reliability of AI-based solutions. Therefore, interdisciplinary solutions combining Big Data, databases of scientific publications, and scientific evidence should be sought and integrated into AI algorithms. Only such an approach has a chance to be considered fully credible and future-oriented based on what will happen in the future, but in particular on the legacy of global science. After all, artificial intelligence works on secondary data and its sphere of thinking similar to the human brain still requires further research, including research into the safety of these solutions for mankind.

The limitations of this scientific work include the lack of fully verified and validated theories and concepts of using this theory, but also comprehensive practical solutions for monitoring and forecasting climate change in all core areas of earth activity and weather changes. Every now and then we hear about the phenomena of weather anomalies causing people's deaths and very large material losses, where the previously used methods of prediction and forecasting have not worked and were not comprehensively included in the systems of mentoring and response of rescue, information and civil protection services of citizens. There is a great deal of work to be done in both science and practice.

12.2 Conclusions

The application of theoretical and practical knowledge in the context of the conditions of AI development against the background of dynamic climate changes requires a new look and the use of numerous modern complex algorithms of conduct and patterns of conduct, including mechanisms combining the use of scientific knowledge, including evidence described in the relevant literature with local conditions and people's experience in given geographical areas. Only very wide access to knowledge, its rapid implementation and experiential learning can improve the effectiveness of people's actions, and in particular the rescue service in responding to climate change. In the business sphere, start-ups and business models should be created that are part of the creation of climate services and the development of business ecosystems, based on which new business models of companies that positively affect climate change and reverse negative trends in weather and climate phenomena will be created.

References

Barends, E., and Rouusseau, D. M. (2018). *Evidence -Based Management, How to use evidence to make better organizational decisions*, Kogan Page Limited, Great Britain and United States.

Jabłoński, M., Jabłoński, A., Janulek, P., Dulęba, D., and Glenszczyk, M. (2025). *Traktat o zasadach zarządzania dowodowego - przyszłość zarządzania* [Treaty on the Principles of Evidence-Based Management - The Future of Management], CeDeWu, Warszawa.

Chapter 13

Recommendations and Strategic Axioms for the Development of Artificial Intelligence in Climate Hazard Management

13.1 Introduction

The purpose of this chapter is to indicate recommendations and strategic axioms for AI development in climate hazard management.

The scope of the chapter includes twenty-four axioms, which are a determinant for further action in the field of science to clarify the laws and rules determining the impact of creating new business models and procedures for saving the climate on climate change.

The subject of the chapter concerns the identification of twenty-four axioms to be further developed in order to extrapolate the knowledge contained herein to further research in this field.

As part of the research and scientific analyses, an analytical and evaluating study was developed, the conclusions of which allowed for the creation of a set of axioms (axioms are statements isolated from all the theorems of a given theory, selected so that all the other theorems of that theory can be derived therefrom. This arrangement of axioms is called axiomatics. The set of axioms and their consequences is an axiomatic system) proving the intensity of theoretical and practical knowledge in the scope of the solved scientific problem related to the issue of AI development in climate hazard management. The defined axioms include the following:

1. Artificial intelligence will radically revolutionize numerous sectors of the economy in the coming years and set new rules and principles for the everyday life of the world's citizens.

DOI: 10.1201/9781003434016-13

2. Artificial intelligence can be used, among others, to build strategies, make decisions based on data analysis and forecast future trends. It is worth remembering, however, that intelligence is originally a property of the human mind. Modern AI systems are based on advanced machine learning algorithms, deep learning and the development of artificial neural networks.
3. The use of artificial intelligence in climate change monitoring processes is significant. This is important in terms of creating new climate services based on the principles of predictive forecasting of weather events, related to agrology, including soil science, sheep farming, poultry industry, etc., forest drainage management and even tourist services.
4. As the climate system is a dynamic and changeable structure, the development of awareness and climate sensitivity among societies in order to effectively protect global climate security is essential.
5. Critical points are distinguished in non-linear climate processes – these are moments when exceeding a certain threshold leads to irreversible and self-reinforcing changes in the climate system. A prerequisite for the effectiveness of the processes of improving the situation pertaining to adverse climate change is attempts to reverse this trend by designing social and business ecosystems focused on revolutionary changes in the behavior of industrialists, business creators and entire societies.
6. The potential of artificial intelligence can be aggregated for the development of so-called climate-smart organizations, i.e., entities that have climate values encoded in their organizational genotype. These values act as a strategic helm, shaping the structure of the organization's business models and strategies, and thus influencing the logic of process and project management in accordance with climate responsibility criteria.
7. Both management and climate sciences are important research areas that, in the context of increasing climate change, have the potential to generate both theoretical and practical solutions. In view of the growing number of climate catastrophes and unpredictable anomalies, these disciplines increasingly overlap. Climate science fosters the understanding of the mechanisms and impact of environmental change, while management sciences provide appropriate concepts and tools to react, adapt and counteract the adverse impact of environmental change.
8. In the face of the worsening climate crisis, the proper targeting of the AI potential towards actions to eliminate and mitigate the impact of

climate change, as well as the construction of early response systems, is becoming a crucial area of activity and strategic decisions.

9. Artificial intelligence opens up wide opportunities for the development of adaptation strategies in response to climate change, at the level of countries, regions and cities, and in the context of individual, responsibly managed climate-sensitive companies.
10. Climate science, as a strongly data-driven field, has been significantly transformed by the era of Big Data and the accompanying technological revolutions. The spectacular successes of Big Data analytics in various sectors over the past decade have also raised high expectations about its effectiveness in addressing the global challenge of climate change. As a result, climate change has become one of the main areas of application of Big Climate Data Analytics.
11. The bridge between the world of plants and animals and technology is a science called bionics. The applications of bionics in the economy and society are extremely inspiring and cover almost all aspects of human existence. Bionics is based on analogous thinking – looking for natural phenomena that can be a reference point in solving specific technical or organizational problems. Such approaches are part of the process of bio-inspiration, which finds effective and efficient application in the development of the concept of artificial intelligence, especially in the context of climate change and changes in natural ecosystems.
12. Artificial intelligence has huge impact potential – both positive and, unfortunately, negative. The intentions of those who use it are crucial. If used in good faith, it can bring huge benefits; however, in the hands of people with bad intentions, it can lead to serious threats on many levels. Therefore, it is vitally important to create a legal and ethical framework to regulate AI development and use. The regulatory system should, on the one hand, protect society from abuse and, on the other hand, not limit the creativity and flexibility of the technologies that constitute its greatest asset.
13. Climate change is a complex and multifaceted issue that can be interpreted differently, often incompletely or incorrectly. There are many false theories and disinformation in public circulation, which significantly hinders a fair debate. Therefore, it is crucial to counteract these phenomena by disseminating knowledge based on scientific cognition. The issue of climate change is unique because it is a classic example of a global externality and involves the problem of managing numerous public goods in conditions of great uncertainty. Effective action against climate change is only possible through international cooperation, as there are currently no effective management mechanisms above the level of nation states.

14. Climate-targeted AI can be an important tool in environmental management processes and in adaptation to climate change. The concept of "climate-creative AI" assumes the use of AI as a conscious and purposeful technological tool to identify and solve specifically defined climate problems. Its operation can be based on a dynamically developed set of so-called climate algorithms, the structure of which contains built-in mechanisms promoting a positive impact on ecosystems. Such algorithms are an attempt to encode the pro-climate logic of operation in the AI architecture, which enables the implementation of sustainable and adaptive solutions in various environmental contexts.
15. A progressive business model, focused on generating positive climate impact, can be seen as a digital adaptation model, shaped not so much by conscious design as by an evolutionary response to growing regulatory and legislative demands. The model uses legal instruments – laws, regulations and directives – as tools to set the decision-making framework in line with dynamically changing climate policy objectives. At the same time, this model does not interfere with current organizational operations and does not require direct management of material resources or industrial infrastructure.
16. Business models integrated with AI technologies represent a new paradigm for business organization, where AI is used to optimize, innovate and create value in new segments of services, products and operations. Their core feature is their ability to process Big Data and apply machine learning methods to generate processes with higher levels of autonomy, flexibility and reactivity. In the context of mass production tailored to the needs of customers, artificial intelligence enables the transformation of classic production models through the implementation of next-generation factories, which are characterized by the ability to independently assess the operating state, dynamic reconfiguration of processes and intelligent operational decision-making.
17. In order to increase the effectiveness of climate action, climate service providers should strive for a systemic understanding of the entire value chain, identifying its core links and actors and engaging in cross-sectoral cooperation. Only by jointly developing services that meet both market needs and social objectives is it possible to create effective and sustainable solutions. In fact, a full understanding of the structure and dynamics of the value chain would benefit all actors, allowing for the better matching of actions and synergistic effects.
18. AI-based technologies should effectively support problem-solving related to the integration of renewable energy sources into the electricity system. This includes forecasting the production of energy

from solar and wind sources, frequency analysis and control of the energy system, as well as assessing the transitional stability of the entire network.

19. The rise of AI and machine learning applications in agricultural sciences is contributing to the transformation of traditional practices towards precise, automated agriculture. These technologies make it possible to predict yields, detect plant diseases, monitor soil and climate conditions, and support decision-making in farm management. However, many AI models are so-called "black boxes", which means a lack of transparency of decision-making mechanisms and limited ability to interpret results. At the same time, artificial intelligence is becoming a core element in emerging digital environmental management systems, including the protection of marine biodiversity, where it supports species identification, risk analysis and the design of conservation strategies based on satellite, sensory and biological data.
20. Climate change, including increased average temperatures, the destabilization of ecosystems and the exacerbation of extreme weather events, contributes to the deterioration of air quality, the increased risk of spreading infectious diseases and an increased health burden associated with thermal shock, cardiovascular diseases and stress. Many health effects of climate change are indirect and seemingly not directly related to temperature, making it difficult to unambiguously identify and attribute them causally. The aim should be to minimize these situations in a systemic way by making evolutionary and revolutionary changes in business ecosystems and changes in citizens' consumption behavior.
21. The integration of AI technologies with satellite observation systems plays an important role in the detection of pollution and environmental anomalies. AI algorithms that process satellite images enable the identification of spatial and temporal deviations that may indicate adverse effects on the climate or illegal environmental activities. The implementation of advanced tools for monitoring and measuring global warming with the use of AI is a key factor for effective climate management. The greenhouse effect is caused by the presence of specific gases in the troposphere, primarily water vapor, methane (CH4), carbon dioxide (CO2) and other greenhouse gases that trap heat radiation, leading to an increase in the Earth's average temperature.
22. In the context of environmental impact management, carbon footprint verification is an important element of sustainability assessment and can be significantly improved through the integration of AI tools. The carbon product footprint of a product or service (CPF) refers

to the total greenhouse gas (CO_2e) emissions generated throughout the life cycle of a product, from the raw material acquisition phase, through production and logistics, to the final waste management phase. The CPF is one of the fundamental indicators included in non-financial ESG reporting, according to the EU Taxonomy. AI technologies support the automation of environmental analysis and the management of complex supply chains, enabling precise emission tracking and efficient management of production processes involving intelligent robots and decision-making systems.

23. The climate transformation, often referred to as Just Transition, is the idea that the pursuit of climate neutrality should be socially just, respecting the rights of workers, local communities and vulnerable groups. The Just Transition Mechanism aims to ensure that no one is left behind in the process of systemic change. At the same time, the climate transition is closely linked to the on-going digital transformation, a phenomenon that, together with the decarbonization process, creates the concept of the Dual Transition, requiring a coordinated and synergistic approach at the level of public policies and governance models.
24. In the context of the digital–climate transition, operational cybersecurity and cyber protection systems play a fundamental role in ensuring the continuity and reliability of environmental management processes. Cybersecurity refers to the implementation of policies and controls as part of operational digital solutions, while cyber protection includes comprehensively securing ICT infrastructure, including critical infrastructure, against cyber-threat incidents – such as ransomware, hacking or cyber-terrorism. The integration of advanced safeguards into climate and energy data management systems is now one of the pillars of the resilience of modern, sustainable economies.

Such a broad and interdisciplinary approach to the axioms developed in the course of research confirms how broad, interdisciplinary and complex the issue of identifying the place and role of artificial intelligence in monitoring and searching for solutions to transform the current situation is, i.e., the increasingly frequent adverse climate impact, including weather anomalies and changes in many areas of activity of our planet Earth towards greater stability and the ability to build action strategies resistant to these phenomena.

13.2 Conclusions

The content of this monograph satisfies the needs of both theoreticians and practitioners in finding ways to determine new directions of scientific research and design new mechanisms for the application of artificial intelligence and technical organizational and conceptual solutions based on the knowledge and experience of machine learning and other solutions constituting the achievements of science over the last few decades. They fit into the normal life of the world's citizens and the modern development of economies, including the development of services based on modern business ecosystems enhancing people's lives and developing new intelligent forms of doing business. Climate change is a major challenge for the whole world. The problem of human migration to places where the climate is more friendly, due, among others, to the high level of air temperatures, the melting of glaciers and changes in the sphere of oceans and seas affect the formation of socio-technical and economic social systems directly and indirectly. Both of these areas functioning at the intersection between politics and the economy significantly affect the creation of the new space for running a business, setting limits but also opportunities that a modern man with a high level of intelligence should implement into the practice of life and business. The subject matter of this book is not easy and requires further scientific research and analysis. However, it is the starting point for the scientific discourse on how to deal with climate change by means of artificial intelligence.

Conclusion

The climate crisis and its incredible dynamics affect the entire spectrum of human existence in the space of society, business and the economy. The degree of emerging climate hazards generates huge human and infrastructural losses. Hence, it becomes crucial to search for technological, organizational, management and operational mechanisms and solutions, the combined implementation of which can mitigate and neutralize constantly emerging climate hazards and threats. The advancing development of artificial intelligence opens up incredible opportunities to predict climate change, catastrophes, anomalies and dysfunctions, and creates new principles for climate management geared towards scientific potential in the disciplines of management and quality sciences and climate sciences. Artificial intelligence and its intellectual strength create new platforms for providing information about climate hazards and modeling climate processes by creating new climate services and shaping new business models that have a positive impact on the climate, new products and new smart climate values. The essence of using artificial intelligence for the effective and efficient management of climate hazards opens the way to initiating decision-making processes based on the development of artificial intelligence theory in evolutionary terms in relation to the dynamics of organizational, managerial and technical–operational changes of companies, cities and individual countries. Therefore, AI technologies can also be used in the construction of digital business models of companies that have a positive impact on climate change. Artificial intelligence is also essential in managing strategic climate information to develop crisis management principles in dynamic terms. Appropriate and properly applied weather forecasting and modelling of climate change by means of AI solutions is also important. Furthermore, it is important to jointly create the digital and climate transformation based on digital solutions treated as a dynamic process of emerging social and economic changes. AI solutions can also be used to verify the carbon footprint of complete supply chains in international terms. Hence, artificial intelligence, its algorithmization and algocracy create a new dimension of decision-making mechanisms in climate hazard management. It seems that the potential use of artificial intelligence to manage climate hazards has not yet been fully exploited to create intelligent climate values in turn.

DOI: 10.1201/9781003434016-14

Index

D

E

F

G

I

K

M

N

O

P

R

S

T

V

Taylor & Francis
Taylor & Francis Group
http://taylorandfrancis.com

For Product Safety Concerns and Information please contact our EU representative GPSR@taylorandfrancis.com Taylor & Francis Verlag GmbH, Kaufingerstraße 24, 80331 München, Germany